ADVANCES IN
DENDRITIC MACROMOLECULES

Volume 3 • 1996

ADVANCES IN DENDRITIC MACROMOLECULES

Editor: GEORGE R. NEWKOME
Department of Chemistry
University of South Florida
Tampa, Florida

VOLUME 3 • 1996

 JAI PRESS INC.

Greenwich, Connecticut   *London, England*

CONTENTS

LIST OF CONTRIBUTORS

Beatriz Alonso   Departamento de Química Inorgánica
                                Universidad Autónoma de Madrid
                                Madrid, Spain

Vincenzo Balzani   Dipartimento di Chimica
                                Università di Bologna
                                Bologna, Italy

Martin R. Bryce   Department of Chemistry
                                University of Durham
                                Durham, England

Sebastiano Campagna   Dipartimento di Chimica Inorganica e
                                Struttura Molecolare
                                Università di Messina
                                Messina, Italy

Carmen M. Casado   Departamento de Química Inorgánica
                                Universidad Autónoma de Madrid
                                Madrid, Spain

Isabel Cuadrado   Departamento de Química Inorgánica
                                Universidad Autónoma de Madrid
                                Madrid, Spain

Gianfranco Denti   Laboratorio di Chimica Inorganica
                                Università di Pisa
                                Pisa, Italy

Wayne Devonport   IBM Research Division
                                Almaden Research Center
                                San Jose, California

Anders Hult   Department of Polymer Technology
                                 Royal Institute of Technology
                                 Stockholm, Sweden

Henrik Ihre   Department of Polymer Technology
                                 Royal Institute of Technology
                                 Stockholm, Sweden

Mats Johansson   Department of Polymer Technology
                                 Royal Institute of Technology
                                 Stockholm, Sweden

Alberto Juris   Dipartimento di Chimica
                                 Università di Bologna
                                 Bologna, Italy

Francisco Lobete   Departamento de Química Inorgánica
                                 Universidad Autónoma de Madrid
                                 Madrid, Spain

José Losada   Departamento de Ingeniería Química
                                 Industrial
                                 Universidad Politécnica de Madrid
                                 Madrid, Spain

Eva Malmström   Department of Polymer Technology
                                 Royal Institute of Technology
                                 Stockholm, Sweden

Moisés Morán   Departamento de Química Inorgánica
                                 Universidad Autónoma de Madrid
                                 Madrid, Spain

Carmen Pascual   Departamento de Química Inorgánica
                                 Universidad Autónoma de Madrid
                                 Madrid, Spain

Concepció Rovira   Institut de Ciència de Materials de Barcelona
                                 (C.S.I.C.)
                                 Campus de la U.A.B.
                                 Bellaterra, Spain

Daniel Ruiz

Institut de Ciència de Materials de Barcelona
(C.S.I.C.)
Campus de la U.A.B.
Bellaterra, Spain

Scolastica Serroni

Dipartimento di Chimica Inorganica e
Struttura Molecolare
Università di Messina
Messina, Italy

Jaume Veciana

Institut de Ciència de Materials de Barcelona
(C.S.I.C.)
Campus de la U.A.B.
Bellaterra, Spain

Nora Ventosa

Institut de Ciència de Materials de Barcelona
(C.S.I.C.)
Campus de la U.A.B.
Bellaterra, Spain

Margherita Venturi

Dipartimento di Chimica
Università di Bologna
Bologna, Italy

PREFACE

The continued interest in dendritic materials as well as the related hyperbranched polymers has sparked the imagination of researchers in many different areas. The incredible increase in annual publications in this topic is best shown in the Figure; and thus as the number on new building blocks and core molecules proliferate, the structural composition of precise and controlled design will grow to meet the imagination of molecular architects. This review series was initially conceived to cover the synthesis and supramolecular chemistry of dendritic or cascade super-molecules as well as their less perfect hyperbranched cousins.

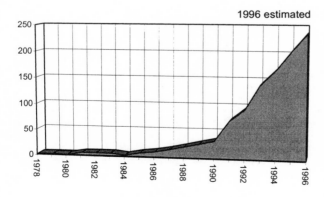

In Chapter 1, Hult and co-workers describe their synthesis and characterization of dendrimers and hyperbranched polyesters, both based on 2,2-bis(hydroxymethyl)propionic acid, as the AB_2-monomer. In Chapter 2, Veciana et al. discuss the advantages and drawbacks of dendritic molecular architectures necessary to create polymeric organic magnetic materials. In Chapter 3, Balzani and colleagues delineate their contributions to the field of polynuclear transition metal complexes in the design and construction of dendritic nanostructures; these luminesence and redox-active complexes suggest their role as photochemical molecular devices operating by photoinduced energy and electron transfer processes. In Chapter 4, Bryce and Devonport review the overall progress on redox-active dendrimers, especially as redox catalysts, organic conductors, modified electrodes, and models for electron transfer proteins. Last but not least in Chapter 5, Cuadrado and collaborators summarize the pioneering research in organometallic dendritic macromolecules and then delineate the redox properties of a series of silicon-based ferrocenyl-containing dendrimers.

I personally wish to thank these authors for their time and effort to complete their contributions to this series and wish them continued success in their further endeavor in the field. Future volumes in this series will continue to highlight the research efforts of others in the field of cascade/dendritic and related hyperbranched macromolecules.

George R. Newkome
Editor

DENDRIMERS AND HYPERBRANCHED ALIPHATIC POLYESTERS BASED ON 2,2-Bis(HYDROXYMETHYL)PROPIONIC ACID (Bis-MPA)

Henrik Ihre, Mats Johansson, Eva Malmström, and Anders Hult

Advances in Dendritic Macromolecules
Volume 3, pages 1–25
Copyright © 1996 by JAI Press Inc.
All rights of reproduction in any form reserved.
ISBN: 0-7623-0069-8

ABSTRACT

The synthesis and characterization of dendrimers and hyperbranched aliphatic poly-esters based on 2,2-bis(hydroxymethyl)propionic acid (bis-MPA) as AB_2-monomer are described. The dendrimers are characterized with NMR and SEC but also with diffusion-NMR to determine their hydrodynamic radii. A new method using a pseudo-stepwise procedure for the synthesis of hydroxyfunctional hyperbranched polyesters using a core molecule yielding hyperbranched polyesters with a high degree of branching is presented. The size of the polyesters is determined by the core:AB_2–monomer ratio. NMR studies using model compounds are employed in order to determine the degree of branching. The hyperbranched polyesters are modified with a number of different end groups, both reactive and nonreactive. The effect of size and end-group structure on the polyesters are beside the chemical characterization studied with respect to the physical and rheological properties. Acrylate functional polyesters are described with respect to curing performance, residual unsaturation, and final film properties.

I. BACKGROUND

In an earlier study, starbranched resins were studied.[1] These resins were interesting in that a twofold increase in molar mass resulted in a lower increase in viscosity than expected from data obtained for corresponding linear polymers. The idea was then to synthesize a hyperbranched polymer based on 2,2-bis(hy-droxymethyl)propionic acid (bis-MPA). Hyperbranched polymers had recently been presented by Fréchet et al.[2] and by Kim et al.[3] The architecture of the hyperbranched polyester somewhat resembled the one of dendrimers, an area pioneered by Tomalia et al.[4] and Newkome et al.[5] half a decade earlier. In our first attempts, pure bis-MPA was esterfied via an acid-catalyzed polyesterification reaction yielding an insoluble polymer. A core molecule was used in the synthesis of dendrimers and we saw no reason why this would not be useful in the synthesis of the hyperbranched structures as well. By copolymerizing bis-MPA with a polyol core a more narrow distribution in molar mass was obtained and the risk of gelation almost eliminated. Since these systems were very promising from several points of view another research project was started, the synthesis of perfect dendrimers based on the same repeating unit, bis-MPA. The aim of these research activities was to gain more understanding on how these highly branched polymers behave and also to compare aliphatic dendrimers and hyperbranched polymers based on the same structure.

II. INTRODUCTION

Dendrimers and hyperbranched polymers are two groups of materials resembling each other. The architectural difference is that dendrimers are perfectly branched structures, while hyperbranched polymers contain defects. Dendrimers are monodispersed while hyperbranched polymers are more dispersed which can be an advantage in some applications.

A manifold of dendrimers have been presented in the literature ranging from polyamidoamine,[6] poly(propylene imine),[7] aromatic polyether[8] and polyester,[9] aliphatic polyether[10] and polyester,[11] polyalkane,[12] polyphenylene,[13] polysilane,[14] and phosphorus[15] dendrimers. Combinations of different backbones as well as architectural modifications have also been presented. For example, the incorporation of chirality in dendrimers,[16] copolymers of linear blocks with dendrimer segments (dendrons),[17] and block copolymers of different dendrons[18] has been described. Numerous applications have been proposed for dendrimers such as biotemplates, liquid membranes, catalysts, or in medical applications.[20]

The synthesis and characterization of a large number of different dendrimers[19,20,21] have been presented in literature. Two different approaches have been employed in the synthesis of dendrimers by convergent[8a,18] or divergent[4,6a] growth. The divergent and convergent growth approaches are two different stepwise procedures in the synthesis of well-defined dendrimers. In the divergent growth, generation by generation of AB_x monomers are added outwards in layers, around a polyfunctional core molecule, until the wanted molar mass/molecular size is obtained. Even at low generations, a large number of functional end groups will be present at the surface of the molecule. This fact may cause problems in the formation of new, fully substituted generations (layers). Purification could also be a problem due to the small differences between the by-products, not fully substituted generations. In the convergent growth approach, dendrons are synthesized from the AB_x monomer and, in a final step, the dendrons are coupled to the polyfunctional core molecule to form a dendrimer of the corresponding generation. The convergent growth approach, developed by Fréchet et al.,[8a] provides a method where each step in the synthesis of the final dendritic structure could be very well controlled since fewer functional groups are involved in each reaction. This should be compared to the rapidly increasing number of functional groups B to be reacted in the divergent approach. The large differences in molar mass between products and by-products in the convergent approach also makes purification easier.

The dendrimers all have a discrete size, shape, and molar mass. The dendrimers exhibit unique properties such as high solubility in various solvents and increased chemical reactivity compared to their corresponding linear analogs. The intrinsic viscosity as a function of molar mass does not obey the Mark–Houwink–Sakurada equation[22] as otherwise expected from a linear polymer (Figure 1).

Hyperbranched polymers are formed by polymerization of AB_x-monomers as first theoretically discussed by Flory.[23] A wide variety of hyperbranched polymer structures such as aromatic polyethers[24] and polyesters,[25] aliphatic polyesters,[26] polyphenylenes,[27] and aromatic polyamides[28] have been described in the literature. The structure of hyperbranched polymers allows some defects, i.e. the degree of branching (DB) is less than one. The synthesis of hyperbranched polymers can often be simplified compared to the one of dendrimers since it is not necessary to use protection/deprotection steps. The most common synthetic route follows a one-pot procedure[24,25a–d,27] where AB_x-monomers are condensated in the presence of a catalyst. Another method using a core molecule and an AB_x-monomer has been described.[25e,26]

The easier synthesis of hyperbranched polymers makes it possible to produce them on a large scale at a reasonable cost giving them an advantage over dendrimers with respect to their use in applications consuming large amounts of material. The properties of hyperbranched polymers are between those of dendrimers and linear polymers, e.g. they exhibit an increased solubility relative to linear analogs but less than dendrimers. The majority of the dendrimers and hyperbranched polymers described are amorphous, but hyperbranched structures exhibiting order such as liquid crystallinity have been presented.[29] A number of material-consuming applications, such as rheology modifiers[27] and coating materials,[26a,39] have been presented where hyperbranched polymers can find use.

The first part of this review deals with the synthesis and characterization of dendritic aliphatic polyesters based on 2,2-bis(dimethylol)propionic acid (bis-MPA) and 1,1,1-tris(hydroxyphenyl)ethane (THPE), as core molecule. The convergent growth approach described earlier was applied. We also present ^{1}H NMR self-diffusion studies of dendrimers of first, second, and third generation.

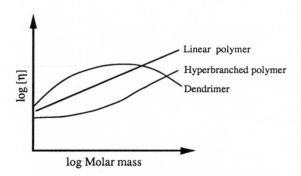

Figure 1. General behavior of the intrinsic viscosity as a function of molar mass for dendrimers, linear, and hyperbranched polymers.

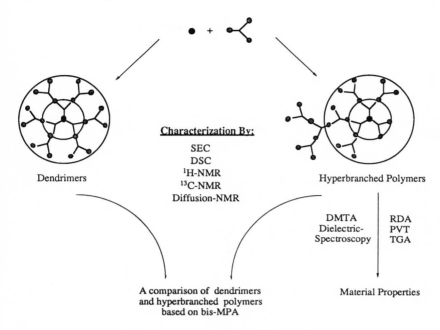

Figure 2. General outline of this review.

In the second part of this review, we report the synthesis of dendrimers and hyperbranched polyesters based on 2,2-bis(methylol)-propionic acid. An extensive study has also been made on the material properties of the hyperbranched polyesters both as thermoplastics and as cross-linkable resins. All characterization techniques and apparatus that have been used throughout this work are described in the Appendix. An outline of the material covered is shown in Figure 2.

III. DENDRIMERS BASED ON Bis-MPA

A. Synthesis

The chosen AB_2 monomer for the synthesis of the dendritic aliphatic polyesters **D1**, **D2**, and **D3** was bis-MPA. In the convergent growth approach, dendrons of certain generations were initially synthesized.[30] In a final step these dendrons were coupled to the polyfunctional core molecule (see Scheme 2). To get an acceptable overall yield it is important that all reactions such as coupling, protection, and deprotection are selective and proceed in high yields since a large number of steps are involved in the synthesis of the final dendrimers.

Table 1. Description of Dendrimers **D1–D3**[a]

Polymer No.	Ratio Core:bis-MPA	Core Molecule	No. of Terminal Groups	Terminal Groups
D1	1:3	THPE	6	actetate
D2	1:9	THPE	12	actetate
D3	1:21	THPE	24	actetate

Note: [a]Abbreviations: 1,1,1-tris(hydroxy-phenyl)ethane (THPE), 2,2-bis(dimethylol)propionic acid (bis-MPA).

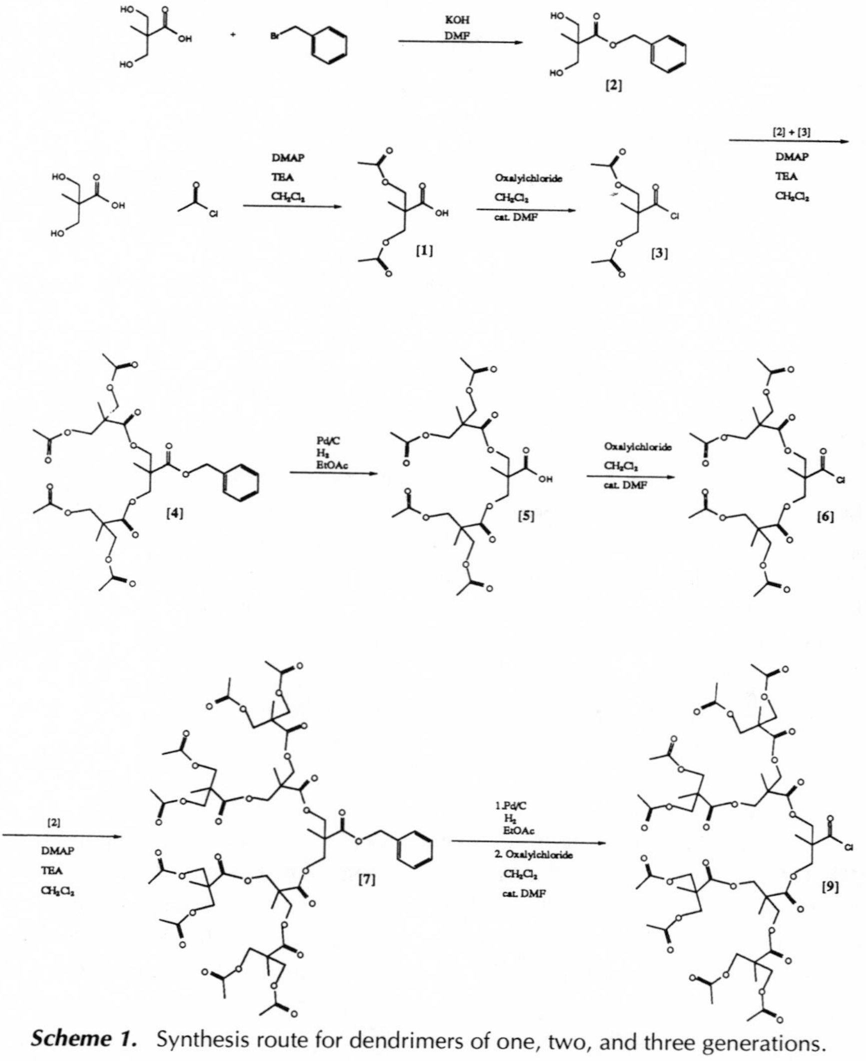

Scheme 1. Synthesis route for dendrimers of one, two, and three generations.

A system of protective groups for the acid and hydroxyl groups in bis-MPA had to be established in order to avoid undesired self-condensation of bis-MPA (Scheme 1). The hydroxyl groups in bis-MPA were protected or deactivated by conversion into the corresponding acetate esters. The diacetate ester of bis-MPA **1** was prepared by reacting bis-MPA with acetylchloride. The benzyl ester group was used to protect the acid group in bis-MPA. The benzyl ester of bis-MPA **2** was prepared by first forming the potassium salt of bis-MPA and in a second step reacting the salt with benzylbromide. The benzyl ester group could selectively be removed in high yields by catalytic hydrogenolysis[31] without affecting the ester bonds formed in the convergent growth. The acid chlorides were prepared by reacting the corresponding acid with oxalyl chloride in CH_2Cl_2, using *N,N*-dimethyl-formamide (DMF) as a catalyst. All esterifications were performed by first converting the acid into the corresponding acid chloride. The ester bond was then formed by reacting a small

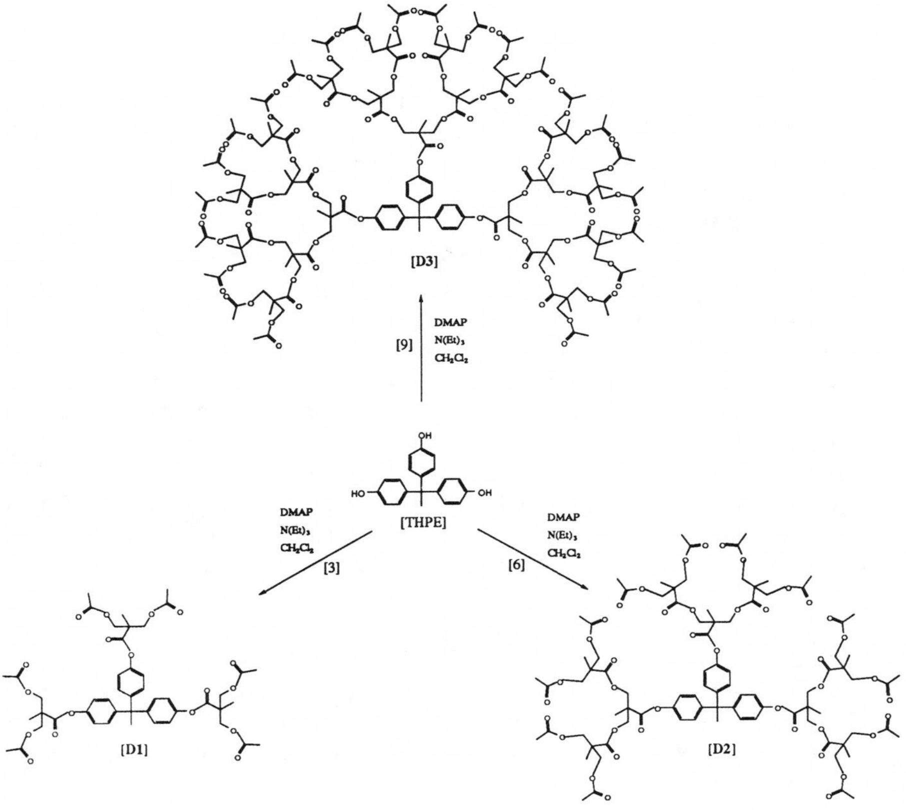

Scheme 2. Synthesis route of dendrons of one, two, and three generations.

excess of the acid chloride with the hydroxyl groups in the presence of triethylamine (TEA) and catalytic amounts of dimethylaminopyridine (DMAP) in CH_2Cl_2. 1,1,1-Tris(hydroxyphenyl)ethane (THPE) was used as core moiety in the formation of the final dendrimers.

The diacetate of bis-MPA **1** was converted into the corresponding acid chloride **3** (Scheme 1). Reaction of acid chloride **3** with the benzylester of bis-MPA **2** according to the standard esterification procedure gave the second generation dendron **4**. Removal of the benzyl ester group by catalytic hydrogenolysis gave the acid **5**. After conversion into the corresponding acid chloride **6** and esterification with the benzylester of bis-MPA **2** the third generation dendron **7** was obtained. Product **7** was deprotected by catalytic hydrogenolysis and converted into the acid chloride **9**.

The dendrimers of one, two, and three generations (**D1**, **D2**, and **D3**) were synthesized by coupling the corresponding dendrons (**3**, **6**, and **9**) with the core molecule (THPE) (Scheme 2).

B. Characterization of the Aliphatic Dendrimers

The techniques used for the characterization of the dendrimers **D1**, **D2**, and **D3** were ^{1}H NMR, ^{13}C NMR, and SEC. Self-diffusion studies were made by pulsed-field spin echo ^{1}H NMR. The simplicity of the ^{1}H NMR and ^{13}C NMR spectra indicates high purity of **D1**, **D2**, and **D3**.

The ^{1}H NMR spectra (Figure 3) clearly distinguished the different generations of final dendrimers. The resonances from the methylene groups in the third generation dendrimer **D3** [4.18, 4.28, and 4.41 ppm (B, B', and B'')], and the methyl groups [1.21, 1.27, and 1.44 ppm (A, A', and A'')], all result in singlets but of significantly different chemical shifts depending on what generation they originate from. The symmetry of the two doublets emanating from the core molecule at 6.96–7.13 ppm (C and D) indicates a fully substituted core molecule. When only one or two dendrons were coupled to the core molecule no symmetry was observed. Clear differences could also be seen between different generations in the ^{13}C NMR spectra of structures **D1**, **D2**, and **D3**. Different chemical shifts were observed for the quaternary carbons in the third generation dendrimer **D3** (46.23, 46.65, and 46.88 ppm), and for the carbonyl carbons (170.75, 171.45, and 171.95 ppm).

Since no adequate SEC standards were available, linear polystyrene was used as standard. As expected, the determined molar masses were not in agreement with the theoretical molar masses. This could be explained by the differences in hydrodynamic volume between linear polystyrene standards and the dendritic polyesters. SEC analyses showed polydispersity values (M_w/M_n) below 1.02 for dendrimers **D1**, **D2**, and **D3**, which was the maximum resolution of the column (Table 2, Figure 4).

From ^{1}H NMR self-diffusion measurements, the molecular diffusion coefficients could be calculated.[32] Since the diffusion coefficients depend on the size and

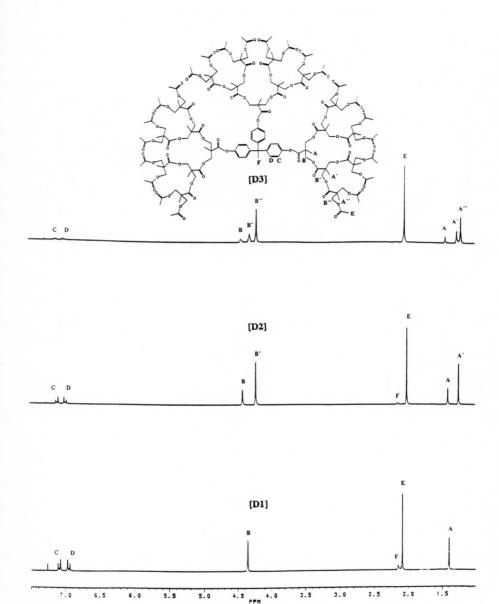

Figure 3. 250-MHz ^{1}H NMR spectra in CDCl$_3$ of dendrimers **D1**, **D2**, and **D3**.

Table 2. SEC Measurements of Dendritic Polyesters of First, Second, and Third Generation (**D1**, **D2**, and **D3**)

Dendrimer	$Mw\ (g\ mol^{-1})$ Theoretical	$Mw\ (g\ mol^{-1})\ SEC$ Data	$Mn\ (g\ mol^{-1})\ SEC$ Data	Mw/Mn SEC Data
D1	906	1386	1365	1.003
D2	1854	1627	1622	1.004
D3	3444	1817	1801	1.009

geometry of the diffusion molecules, this method may be used as a tool for estimating the molecular sizes.[33] In this work, we present measurements that yield the diffusion coefficients and estimations of the molecular sizes.

The self-diffusion coefficients were calculated, and the effective radii of the dendrimers were estimated from the diffusion coefficients by assuming a spherical geometry for all dendrimers. The so-obtained radii were 7.8, 10.3, and 12.6 Å for the first, second, and third generation dendrimer, respectively.

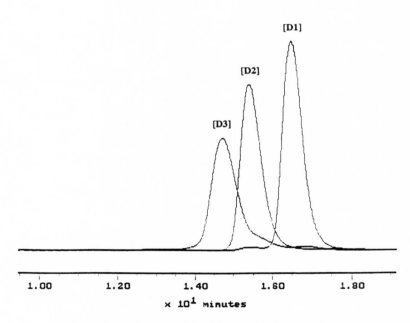

Figure 4. SEC-traces of dendrimers **D1**, **D2**, and **D3**.

C. Summary Dendrimers

The convergent growth approach showed to be a useful method in the synthesis of monodisperse dendrimers. The use of bis-MPA, as a building block, resulted in dendrimers with simple ^{1}H NMR and ^{13}C NMR spectra where different generations could be separated from each other.

Molecular self-diffusion studies by pulsed-field spin echo ^{1}H NMR turned out to be a convenient method to estimate the hydrodynamic radii of dendrimers of different generations.

Due to the multistep synthesis of the dendrimers, it was difficult to produce sample amounts enough for mechanical testing. This is a recurrent problem when dealing with dendrimers and is probably the reason why almost no material properties are reported in the literature. Our intentions were to synthesize dendrimers in sufficient amounts so they could be used as reference substances to compare their material properties with those of the hyperbranched polymers.

IV. HYPERBRANCHED POLYESTERS BASED ON Bis-MPA

A. Synthesis

Synthesis of the Hydroxyfunctional Base Polyester

Hydroxyfunctional hyperbranched polyesters were synthesized in the melt via an acid-catalyzed polyesterification reaction. The polyesterifications were carried out in a pseudo-one-step manner. The acid catalyst, the core moiety, and bis-MPA were added to a glass vessel in a ratio according to the stoichiometry of the first generation. The vessel was placed in a preheated oil bath. After approximately 3 hours, under argon purging and a vacuum, alternately bis-MPA, corresponding to the stoichiometric amount of the second generation, and more acid catalyst were added to the vessel. This reaction procedure was repeated until the desired molar ratio of core molecule:bis-MPA was obtained. A schematic representation of the synthesis of a hyperbranched polyester (ratio TMP:bis-MPA = 1:21) is outlined in Figure 5. All hyperbranched polyesters used in this survey are described in Table 3.

Since it is difficult to determine the molar mass of these polymers, we have chosen to refer to the stoichiometric ratio of core molecule to bis-MPA. This will be further discussed in Section (IV.B).

Synthesis of Differently End-Capped Polyesters

The resulting base polyester has numerous hydroxyl groups, as terminal units. In order to make them cross-linkable (thermosets) or to investigate how different end-groups affect the properties several different end-cappings have been used (Figure 6). The end-capping reactions have most frequently been done with the corresponding acid chlorides.[34]

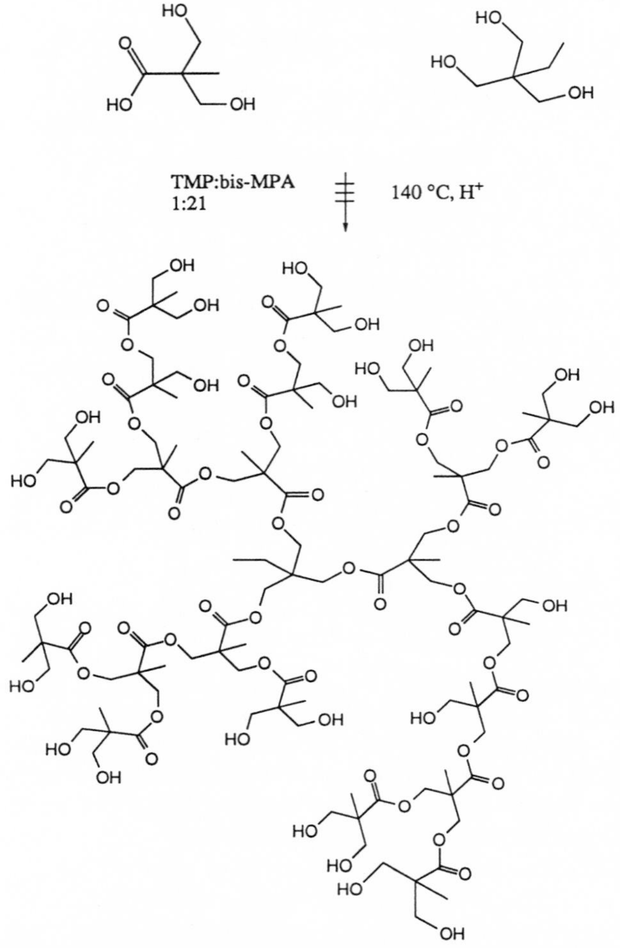

Figure 5. A schematic representation of the synthesis of a hydroxyfunctional base-polyester with ratio TMP/bis-MPA 1:21 with a degree of branching of around 80%.

B. Characterization of Aliphatic Hyperbranched Polyesters

Molar Mass

The architecture of hyperbranched polymers and dendrimers is connected with difficulties in determining molar mass. Many of the common characterization techniques—e.g. size exclusion chromatography (SEC)—used for polymers are relative methods where polymer standards of known molar mass and dispersity are needed for calibration. Highly branched polymers exhibit a different relationship between molar mass and hydrodynamic radius than their linear counterparts.

Table 3. Description of the Hyperbranched Polyesters **H1–H15**[a]

Polymer No.	Ratio Core:bis-MPA	Core Molecule	No. of Terminal Groups	Terminal Groups
H1	1:9	TMP	12	hydroxyl
H2	1:21	TMP	24	hydroxyl
H3	1:45	TMP	48	hydroxyl
H4	1:93	TMP	96	hydroxyl
H5	1:189	TMP	192	hydroxyl
H6	1:381	TMP	384	hydroxyl
H7	1:45	TMP	48	propionate
H8	1:45	TMP	48	benzoate
H9	1:93	TMP	96	acetate
H10	1:93	TMP	96	benzoate
H11	1:21	TMP	24	propionate:acrylate (1:1)
H12	1:21	TMP	24	benzoate:acrylate (1:1)
H13	1:21	TMP	24	hydroxyl:acrylate (1:1)
H14	1:765	TMP	768	silyl ether
H15	1:1533	TMP	1536	hydroxyl

Note: [a]Abbreviations: 2-ethyl-2-(hydroxymethyl)-1,3-propanediol (TMP), 2,2-bis(hydroxymethyl) propionic acid (bis-MPA).

Because there are no appropriate calibration standards for hyperbranched polymers, the results from SEC-measurements are difficult to interpret. The hydrodynamic radius is also strongly affected by the solvent used in the analyses,[33] and this can further complicate the results. Another problem with SEC-measurements is that the

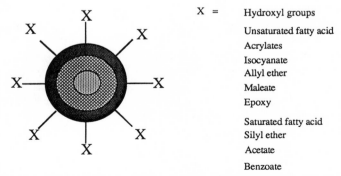

X =	Hydroxyl groups
	Unsaturated fatty acid
	Acrylates
	Isocyanate
	Allyl ether
	Maleate
	Epoxy
	Saturated fatty acid
	Silyl ether
	Acetate
	Benzoate

Figure 6. A schematic representation of different end-cappings that have been synthesized.

Table 4. Results Obtained When Polyesters **H1–H6** Were Characterized with SEC and Their Degree of Branching as Determined by ^{13}C-NMR

Polymer No.	Theor. Molar Mass	SEC[a]			Degree of Branching (%)
		M_w (g mol^{-1})	M_n (g mol^{-1})	M_w/M_n	
H1	1 179	1 903	1 400	1.36	96
H2	2 573	2 616	1 881	1.39	92
H3	5 359	4 472	3 052	1.47	87
H4	10 933	6 905	4 274	1.62	83
H5	22 080	8 545	4 565	1.87	83
H6	44 374	10 765	5 598	1.92	83

Note: [a]As calibrated against linear polystyrene standards.

hydroxyl-terminated samples interact with the column material. It has been claimed that the column packing can be ruined by this interaction.[35]

All base polyesters have been characterized with SEC (Table 4). The polydispersity indices indicate a narrow distribution of hydrodynamic radius. Three differently end-capped polyesters, **H4**, **H9**, and **H10**, were analyzed in THF and acetone. The results were extremely sensitive to the change in polarity of the solvent even though the polydispersity indices agreed quite well.

The dendrimers described in Section III of this chapter were used as calibration standards for the SEC measurements. Sample **H9** was subsequently analyzed, and this measurement indicated a higher dispersity than measurements using linear standards. It is noteworthy that the dendrimers used for calibration had a different core and were quite small even though it is not certain that a dendritic shape (spherical) has been reached.

Some introductory solid state NMR experiments were performed on three hydroxyl-terminated hyperbranched polyesters: **H2**, **H5**, and **H15**. The results are difficult to interpret. One-pulse solid state ^{13}C-NMR (in this experiment only the mobile groups are detectable; Figure 7) indicates that the mobility within the polyester skeleton is greatly diminished as the ratio TMP:bis-MPA is increased. The mobility of the methyl group in the repeating unit was maintained going from **H2** to **H5**, while the mobility of the quaternary carbon was reduced at the sixth generation and totally frozen in at the ninth generation. These results suggested a more crowded structure for higher TMP:bis-MPA ratios and thus indicate a higher molar mass.

Degree of Branching

One of the most characteristic features of a hyperbranched polymer is its degree of branching. The degree of branching (DB) has been defined by Fréchet et al.[2] as:

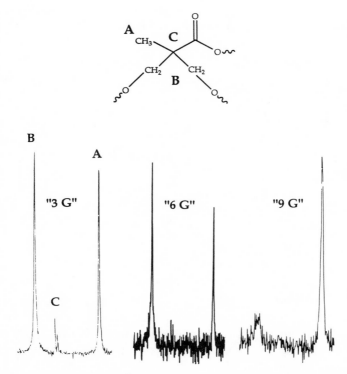

Figure 7. One-pulse solid-state ^{13}C NMR spectra of **H2**, **H5**, and **H15** (10–80 ppm). The disappearance of the –CH$_2$–peak indicates that the mobility is greatly reduced when the ratio TMP:bis-MPA is increased.

$$DB\ (\%) = \frac{\Sigma(\text{terminal units}) + \Sigma(\text{dendritic units})}{\Sigma(\text{all units})} \times 100$$

The degree of branching for a dendrimer is always equal to one, but for a hyperbranched polymer it is always less than one.

A closer investigation of the hyperbranched base polyester reveals that the repeating unit can be incorporated in three different ways in the polymer skeleton, assuming that no side reactions occur. The terminal units all have both their hydroxyl groups left unreacted (C), while bis-MPA in interior layers are either fully branched (A) or linearly incorporated (B) (Figure 8). These linearly incorporated building blocks do not exist in a dendrimer. In order to mimic these differently incorporated building blocks, low molar mass model compounds were synthesized and characterized with ^{13}C-NMR (Table 5). The base polyesters were subsequently

Table 5. The Low Molar Mass Model Compounds that were Synthesized in Order to Determine the Degree of Branching of the Hyperbranched Polyesters

	^{1}H-NMR (ppm)		^{13}C-NMR (ppm)		
Model Compound	$H_3C-C\overset{/}{\underset{\backslash}{-}}$	$-CH_2-\langle$	$-\overset{/}{\underset{\backslash}{C}}-$	$-CH_2-\langle$	$H_3\underline{C}-C\overset{/}{\underset{\backslash}{-}}$
HO—⟨—OH / COOCH₂CH₃	1.06	3.66 – 3.93 (q)	50.65	60.57	14.36
HO—⟨A B—OOCH₃ / COOCH₂CH₃	1.18	A) 3.66 – 3.69 B) 4.18 – 4.35	48.85	A) 60.95 B) 65.13	14.35
H₃CCOO—⟨—OOCCH₃ / COOCH₂CH₃	1.12 – 1.28	4.21	46.80	65.82	14.34

analyzed with ^{13}C-NMR. Different building blocks were quantified using the INVGATE pulse-sequence in order to suppress the NOE effect. The degree of branching was calculated from the results and are reported in Table 4.

The degree of branching was found to be high (> 80%) and almost invariable with the ratio TMP:bis-MPA. The degree of branching for our system is higher than for most systems reported in the literature. It is the pseudo-one-step procedure that favors this as the reaction mixture is phase-separated after each addition of bis-MPA. The reason for the phase separation is the different melting temperatures of, in the first step, the polyol and bis-MPA and in the following reaction step, the polyester and bis-MPA. The polyesterification reaction is performed at a reaction temperature that is lower than the melting temperature of bis-MPA. Thus, the pseudo-one-step concept was utilized to favor the reaction between a free bis-MPA monomer and the hyperbranched skeleton instead of the reaction with another free

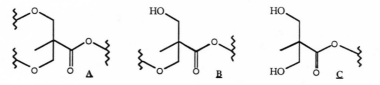

Figure 8. The differently incorporated bis-MPA units in the hyperbranched skeleton, assuming 100% conversion of carboxylic acid groups. **A** is the dendritic unit, **B** the linearly incorporated unit, and **C** the terminal unit.

bis-MPA by keeping the ratio hydroxyl groups (skeleton) to hydroxyl groups (bis-MPA) as high as possible. Further studies with different polyol cores are in progress and will be reported later.

It is difficult to assess the conversion of the hydroxyl groups on the core since the percentage contribution from core atoms diminish rapidly when the samples are analyzed, with e.g. NMR. However, when bis-MPA was esterified in absence of core moiety the resulting polymer was very difficult to dissolve in any solvent. This serves as indirect evidence of the importance of the core moiety.

Size

Hydroxyfunctional base polyesters were treated with hexamethylene disilazane and yielded fully silylated hyperbranched polyesters that provided enough contrast to be analyzed with transmission electron microscopy (TEM). Samples were prepared by casting a solution containing the fully silylated dendrimer and poly(methyl methacrylate) on a grid. Samples were dried in a vacuum oven before being analyzed. As the PMMA was burned down by the electron beam, the silylated hyperbranched polyester, **H14**, became visible (Figure 9). However, the formation of agglomerates are impossible to prevent with this sample preparation technique, which also can be seen in Figure 9.

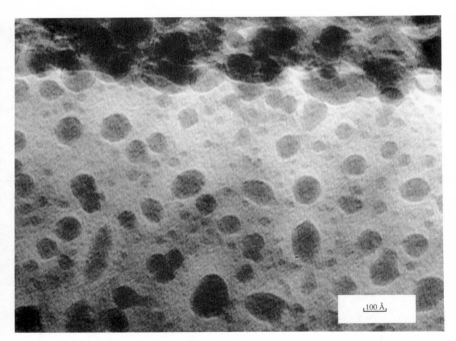

Figure 9. TEM-picture of a fully silylated hyperbranched polyester, **H14**.

C. Material Properties

The hydroxyfunctional hyperbranched polyesters have been characterized with respect to their mechanical and rheological properties, both as thermoplastics and in cross-linked networks. The high number of terminal groups in hyperbranched polymers has a large impact on the properties, and also makes it easy to functionalize the polymers for various applications. One option is to attach reactive groups at chain ends, forming a cross-linkable polymer. Variations in functionality and the type of functional groups will affect both the polymer properties and the final cross-linked material properties.

Mechanical and Physical Properties

All the thermoplastic hyperbranched polyesters are transparent, slightly yellow solids below and viscous liquids above the glass transition temperature, T_g. Thermogravimetric (TGA) studies on a hydroxyfunctional hyperbranched polyester, **H6**, show that the thermal stability is good. The weight loss was only 3.5 wt% up to 340 °C, where the thermal degradation started (Figure 10).

The density of the hydroxyfunctional hyperbranched polyesters is 1.295 g cm^{-3} and pressure–volume–temperature (PVT) measurements show that the thermal expansion and compressibility are slightly lower compared to polar linear polymers, such as PVC, poly(ε-caprolactone), and poly(epichlorohydrine).[37]

Dielectric spectroscopy was performed on **H4, H9**, and **H10**,[36] all based on the same base polyester but with different end groups. To obtain films, samples were mixed 50/50 by weight with dielectrically inactive polyethylene. Sample **H4**, having hydroxyl groups as end groups, exhibited a glass transition at 30 °C, while the glass transition for **H9** (aliphatic) was found at 6 °C and for **H10** (aromatic) at 52 °C. This showed that the interactions between the end groups determines the

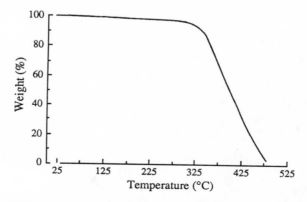

Figure 10. TGA measurement on a hydroxyfunctional hyperbranched polyester (**H6**). The scan was performed in nitrogen atmosphere at a heating rate of 10K min^{-1}.

properties of the polymer to a very high extent. The lowest glass transition was found for the aliphatic terminated sample, having weak interactions, while the glass transition was remarkably increased with aromatic end groups which have an ability to pack the phenylene rings very close together giving strong interactions.

Properties in Melt

A series of different hydroxyfunctional hyperbranched polyesters (**H1–H6**) with increasing ratio TMP:bis-MPA was studied. The tests were made on samples quenched from melt. As discussed previously, the molar masses for these polymers are difficult to determine and the results are therefore presented as a function of the ratio bis-MPA:TMP, which can be directly related to the theoretical molar mass. The complex dynamic viscosity (η^*) of hyperbranched polyesters show an increase in viscosity with size which levels out at a certain value (Figure 11). The corresponding linear polymers would exhibit a linear relationship η^* versus log molar mass and hence have a higher melt viscosity. The hydroxyfunctional polyesters exhibit a Newtonian behavior[37] within a medium shear range (10^{-1}–10^2 rad s^{-1}).

Figure 12 presents η^* as a function of temperature for three hyperbranched polyesters with different end groups. A comparison shows that the properties of hyperbranched polymers are greatly affected by the structure of the terminal groups. The T_g decreases from 35 °C with terminal hydroxyl groups to 15 °C for benzoates and −20 °C for propionates.

Three different acrylates have been synthesized and characterized.[38] The same hydroxy-functional hyperbranched polyester (**H3**) was used as a base for all resins. Approximately 50% of the terminal hydroxyls were acrylated and the rest either

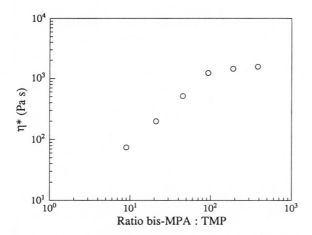

Figure 11. Complex dynamic viscosity (η^*) for hydroxyfunctional hyperbranched polyesters as a function of the ratio bis-MPA:TMP, polyesters, **H1–H6**.

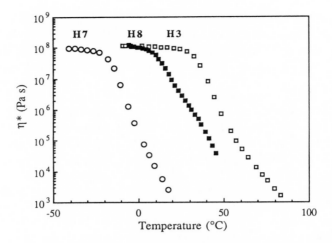

Figure 12. η^* as a function of temperature for hyperbranched polyesters with either hydroxyl (**H3**), benzoate (**H8**), or propionate (**H7**) terminal groups. The ratio TMP:bis-MPA is 1:45 in all polymers, frequency 6.28 rad s^{-1}.

left as hydroxyls (**H13**) or modified to propionates (**H11**) or benzoates (**H12**). All resins were slightly yellow viscous liquids with a difference in dynamic viscosity ranging from ~7.5 MPas for the resin with terminal hydroxyls to 0.75 MPas with benzoates and 0.11 MPas with propionates measured at 8 °C (not at ambient temperature due to experimental difficulties) and 6.28 rad s^{-1}. The T_g for the resins changes with the structure of the terminal units. The softening point is ~–30 °C for the resin with propionate groups, a few degrees higher with aromatic groups, and ~–10 °C with hydroxyl groups as determined by the onset of the drop in η^*. The resins exhibit a Newtonian behavior, i.e. no shear thinning can be observed within a medium shear rate range. The difference in T_g and viscosity is less compared to the thermoplastic polyesters described above since 50% of the end groups are acrylates, hence the variation in structure only concerns 50% of the terminal groups.

Network Structures Based on Hyperbranched Resins

Resins **H11–H13** were UV-cured using a free radical photoinitiator (Irgacure 184™, 1-benzoylcyclohexanol) in a Mini-Cure™ equipped with medium-pressure mercury lamps (80 W/cm). The samples were postbaked at 100 °C for 20 minutes to obtain a similar thermal history for all samples.

All films were touch-dried after one passage through the Mini-Cure™ (i.e. no oxygen inhibition was noticed). Raman spectra of the resins before and after curing showed no detectable amount of residual unsaturation (i.e. less than 5%). This showed that all acrylate groups are accessible to polymerization and not "caged" in the hyperbranched structure.

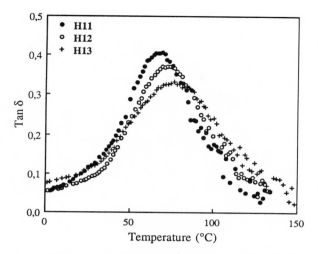

Figure 13. Tan δ as function of temperature for cross-linked films of three different acrylate resins with either hydroxyl (**H13**), benzoate (**H12**), or propionate (**H11**) terminal groups.

Dynamical mechanical measurements on the cured films showed only a small decrease in T_g depending on the terminal units: from 77 °C for the hydroxy, 72 °C for the benzoate, and 67 °C for the propionate functional as determined by the tan δ peak (Figure 13). The cross-linking gives, as expected, a dramatic increase in T_g for each system, but also reduces the differences in T_g as a function of the structure of the terminal groups.

The height and width of the glass transition peak are changed, dependent on the terminal groups. This is indicative of a decreased mobility of the network containing hydroxyls, stronger polar interactions (relative to benzoate functional), and an even larger difference to the propionate functional structure. Since the amount of acrylic groups is the same for all resins, the decreased mobility can be attributed to the mobility of the terminal groups and not to the cross-linking density.

Another resin application based on the same hyperbranched polyester structure described herein is low-VOC alkyds,[39] which have very low viscosity and high reactivity compared to conventional high-solid alkyds. Other resin structures are unsaturated polyesters, polyurethane dispersions, and epoxides.[40]

D. Summary: Hyperbranched Polyesters

A versatile way of synthesizing hydroxyfunctional hyperbranched aliphatic polyesters using 2,2-bis(dimethylol)propionic acid together with a polyol core through a pseudo-step-wise procedure has been presented. The polyesters were shown to have a very high degree of branching (~80%), as determined with

^{13}C-NMR. The use of a core moiety made it possible to synthesize hyperbranched polyesters with varying size, depending on the ratio core:AB_2–monomer. The pseudo-step-wise procedure has been suggested to favor a high degree of branching due to phase-separation phenomena throughout the reaction. A series of different end groups, both reactive and nonreactive, has been attached to the polyesters. The hyperbranched polyesters all exhibited a Newtonian behavior indicating that almost no entanglements between the molecules are present. A comparison of the properties for the thermoplastic hyperbranched polyesters and cross-linked structures showed that the major parameter determining the glass transition for each system is the mobility of the terminal groups. Reduced mobility, either due to cross-linking or polar interactions between the end groups, will both increase the glass transition temperature and yield a harder material.

V. CONCLUSIONS

First, second, and third generation dendrimers based on THPE (core) and bis-MPA (repeating unit) have been synthesized and characterized. SEC and NMR analyses verified that the dendrimers were monodisperse. Diffusion NMR was used to determine the dendrimers' hydrodynamic radii in deuterated chloroform. First, second, and third generation dendrimers were found to be 7.8, 10.3, and 12.6 Å, respectively. These were in agreement with sizes as determined from molecular models (Büchi).

Hydroxyfunctional hyperbranched polyesters based on bis-MPA and a polyol core were synthesized in a pseudo-one-step manner using acid catalyst. The polyesters were found to have a high degree of branching (80%). The thermoplastic polyesters all behaved like Newtonian liquids in a medium shear range. Due to their manifold of end groups, they are easy to functionalize. They were partly acrylated and turned out to be useful in coating applications. Fast curing films with no detectable residual unsaturation left were produced. The viscosity of the resin and the glass transition temperature of the cured films were found to be strongly dependent on the nature of the end groups. With more polar end groups, higher viscosity and glass transition temperatures, respectively, are realized. The manifold of end groups also gave extraordinary possibilities to tailor polymers for certain applications. By varying the nature of end groups, very different polymers can be obtained with the same backbone. The effect of the end groups has still to be further investigated before these possibilities are fully understood.

VI. PRESENT AND FUTURE WORK

Our work with perfect dendrimers continues. The goal is to synthesize at least four generations and then make extensive comparisons between hyperbranched polymers and dendrimers, both from a physical and chemical properties points of view.

A question of major interest is what influence the degree of branching has on the hyperbranched polymers' properties. Several authors suggest that the higher degree of branching for a hyperbranched polymer the better resemblance to a dendrimer. It is of great interest to compare hyperbranched polymers with a different degree of branching to see if there is a discontinuous change in properties at a certain degree of branching. Another area of major importance for hyperbranched polymers is coating applications. It is interesting to further investigate the relation between structure and properties.

VII. APPENDIX: ALL APPARATUS AND TECHNIQUES THAT HAVE BEEN USED THROUGHOUT THIS WORK

Technique	Apparatus	Application
Liquid chromatography (LC)	Columns packed with silica gel 60 (230–400 mesh ASTM); hexane/ethyl acetate gradient	Purification
Infrared spectroscopy (IR)	Perkin-Elmer 1730 FTIR	Structure characterization
^{1}H-NMR	Bruker WP 250 MHz using the solvent peak as a reference peak	Structure characterization
^{1}H- and ^{13}C-NMR	Bruker WP 400 MHz using the solvent as a reference peak	Structure characterization, degree of branching
Dynamic mechanical thermal analysis (DMTA)	Polymer Laboratories DMTA	Mechanical properties E-modulus, cross-link density
Pendulum hardness	König pendulum	Film hardness
Rheology	Rheometrics RDAII equipped with parallel plates of diameter 7.9 or 25 mm	Melt viscosity, rheological behavior
Raman spectroscopy	Perkin-Elmer FT-Raman 1700X	Structure characterization, residual unsaturation
Differential scanning calorimetry (DSC)	Perkin-Elmer DSC7	Thermal transitions
Pressure-volume-temperature (PVT)	Gnomix PVT-apparatus	Compressibility and thermal expansion
Thermal gravimetric analysis (TGA)	Perkin-Elmer TGA7	Thermal stability
Densitometry	Mettler scale using a ME-210250 density determination kit	Density
X-ray diffraction (SAXS)	Statton camera, Cu Ka-radiation; Philips PW 1830 generator	Characterization of structural ordering
Size exclusion chromatography (SEC)	Waters SEC system: WISP 710 automated injector; Water 410 DR μ-Styragel columns 500, 10^5, 10^4, 10^3, 100 Å, solvent:THF	Determination of molar mass and P.D.I.

Technique	Apparatus	Application
Diffusion NMR	See Ref. 30	Hydrodynamic radius in solution
Transmission electron microscopy (TEM)	JEOL JEM 100B	Particle dimension
UV-curing	Mini-Cure™	UV-curing
Dielectric spectroscopy	See Ref. 36	Relaxation phenomena, internal mobility

ACKNOWLEDGMENTS

This work was made possible through a scholarship from Wilhelm Becker AB and financial support from Perstorp Polyols AB, Sweden and by the Swedish Research Council for Engineering Sciences (Dnr: 93-1075) which are gratefully acknowledged. We are also indebted to Dr. Christine Boeffel, Max-Planck Institute, Mainz, for doing the solid-state ^{13}C-NMR measurements and to Ph.D. Erik Söderlind, Department of Chemistry, Physical Chemistry, Royal Institute of Technology for diffusion-NMR measurements.

REFERENCES

1. Johansson, M.; Trollsås, M.; Hult, A. *J. Polym. Sci.: Part A: Polym. Chem.* **1992**, *30*, 2203.
2. Hawker, C. J.; Lee, R.; Fréchet, J. M. J. *J. Am. Chem. Soc.* **1991**, *113*, 4583.
3. Kim, Y. H.; Webster, O. W. *J. Am. Chem. Soc.* **1990**, *112*, 4592.
4. Tomalia, D. A.; Baker, H.; Dewald, J.; Hall, M.; Kallos, G.; Martin, S.; Roeck, J.; Ryder, J.; Smith, P. *Polym. J.* **1985**, *17*, 117.
5. Newkome, G. R.; Yao, Z.; Baker, G. R.; Gupta, V. K. *J. Org. Chem.* **1985**, *50*, 2003.
6. (a) Tomalia, D. A.; Baker, H.; Dewald, J.; Hall, M.; Kallos, G.; Martin, S.; Roeck, J.; Ryder, J.; Smith, P. *Macromolecules* **1986**, *19*, 2466; (b) Tomalia, D. A.; Hedstrand, D. M.; Ferritto, M. S. *Macromolecules* **1991**, *24*, 1435.
7. (a) de Brabander-van den Berg, E. M. M.; Meijer, E. W. *Angew. Chem. Int. Ed. Engl.* **1993**, *32*, 1308; (b) de Brabander-van den Berg, E. M. M.; Nijenhuis, A.; Mure, M.; Keulen, J.; Reintjens, R.; Vandenbooren, F.; Bosman, B.; de Raat, R.; Frijns, T.; v. d. Wal, S.; Castelijns, M.; Put, J.; Meijer, E. W. *Macromol. Symp.* **1994**, *77*, 51.
8. (a) Hawker, C.; Fréchet, J. M. J. *J. Chem. Soc., Chem. Commun.* **1990**, 1010; (b) Wooley, K. L.; Hawker, C. J.; Fréchet, J. M. J. *J. Chem. Soc., Perkin Trans. 1* **1991**, 1059.
9. (a) Miller, T. M.; Kwock, E. W.; Neenan, T. X. *Macromolecules* **1992**, *25*, 3143; (b) Hawker, C. J.; Fréchet, J. M. J. *J. Chem. Soc., Perkin Trans. 1* **1992**, 2459.
10. Buyle Padias, A.; Hall, H. K.; Tomalia, D. A.; McConnell, J. R. *J. Org. Chem.* **1987**, *52*, 5305.
11. Ihre, I.; Hult, A. Presented at the *35th IUPAC International Symposium* in Akron, Ohio, July 11–15, 1994.
12. (a) Newkome, G. R.; Moorefield, C. N.; Baker, G. R.; Johnson, A. L.; Behera, R. K. *Angew. Chem. Int. Ed. Engl.* **1991**, *30*, 1176; (b) Newkome, G. R.; Moorefield, C. N.; Baker, G. R.; Saunders, M. J.; Grossman, S. H. *Angew. Chem. Int. Ed. Engl.* **1991**, *30*, 1178.
13. Miller, T. M.; Neenan, T. X.; Zayas, R.; Bair, H. E. *J. Am. Chem. Soc.* **1992**, *114*, 1018.
14. van der Made, A. W.; van Leeuwen, P. W. N. M. *J. Chem. Soc., Chem. Commun.* **1992**, 1400.
15. Launay, N.; Caminade, A.-M.; Majoral, J.-P. *J. Am. Chem. Soc.* **1995**, *117*, 3282.
16. (a) Lapierre, J.-M.; Skobridis, K.; Seebach, D. *Helv. Chim. Acta* **1993**, *76*, 419; (b) Seebach, D.; Lapierre, J.-M.; Greiveldinger, G.; Skobridis, K. *Helv. Chim. Acta* **1994**, *77*, 1673; (c) Seebach,

D.; Lapierre, J.-M.; Skobridis, K.; Greiveldinger, G. *Angew. Chem. Int. Ed. Engl.* **1994**, *33*, 440; (d) Kremers, J. A.; Meijer, E. W. *J. Org. Chem.* **1994**, *59*, 4262.

17. (a) Gitsov, I.; Wooley, K. L.; Fréchet, J. M. J. *Angew. Chem. Int. Ed. Engl.* **1992**, *31*, 1200; (b) Gitsov, I.; Wooley, K. L.; Hawker, C. J.; Ivanova, P. T.; Fréchet, J. M. J. *Macromolecules* **1993**, *26*, 5621; (c) Gitsov, I.; Fréchet, J. M. J. *Macromolecules* **1993**, *26*, 6536.

18. Hawker, C. J.; Fréchet, J. M. J. *J. Am. Chem. Soc.* **1992**, *114*, 8405.

19. Newkome, G. R.; Moorefield, C. N.; Baker, G. R. *Aldrichim. Acta* **1992**, *25*, 31.

20. Tomalia, D. A.; Naylor, A. M.; Goddard III, W. A. *Angew. Chem. Int. Ed. Engl.* **1990**, *29*, 138.

21. Fréchet, J. M. J. *Science* **1994**, *263*, 1710.

22. Mourey, T. H.; Turner, S. R.; Rubenstein, M.; Fréchet J. M. J.; Hawker, C. J.; Wooley, K. L. *Macromolecules* **1992**, *25*, 2401.

23. Flory, P. J. *J. Am. Chem. Soc.* **1952**, *74*, 2718.

24. Chu, F.; Hawker, C. J. *Polym. Bull.* **1993**, *30*, 265.

25. (a) Turner, S. R.; Voit, B. L.; Mourey, T. H. *Macromolecules* **1993**, *26*, 4617; (b) Turner, S. R.; Walter, F.; Voit, B. I.; Mourey, T. H. *Macromolecules* **1994**, *27*, 1611; (c) Kricheldorf, H. R.; Stöber, O. *Macromol. Rapid Commun.* **1994**, *15*, 87; (d) Wooley, K. L.; Hawker, C. J.; Lee, R.; Fréchet, J. M. J. *Polym. J.* **1994**, *26*, 187; (e) Feast, W. J.; Stainton, N. M. *J. Mater. Chem.* **1995**, *5*, 405.

26. (a) Johansson, M.; Malmström, E.; Hult, A. *J. Polym. Sci.: Part A: Chem. Ed.* **1993**, *31*, 619; (b) Malmström, E.; Johansson, M.; Hult, A. *Macromolecules* **1995**, *28*, 1698.

27. Kim, Y. H.; Webster, O. W. *Macromolecules* **1992**, *25*, 5561.

28. Kim, Y. H. *Macromol. Symp.* **1994**, *77*, 21.

29. (a) Percec, V.; Kawasumi, M. *Macromolecules* **1992**, *25*, 3843; (b) Percec, V.; Cho, C. G.; Pugh, C.; Tomazos, D. *Macromolecules* **1992**, *25*, 1164.

30. Ihre, H.; Hult, A.; Söderlind, E. *J. Am. Chem. Soc.* **1996**, *118*, 6388.

31. Vogel, A. I. *Vogels's Textbook of Practical Organic Chemistry*; Longman Scientific & Technical: Essex, 1989.

32. Söderman, O.; Stilbs, P. *Prog. Nucl. Magn. Reson. Spectrosc.* **1994**, 26, 445.

33. Young, J. K.; Baker, G. R.; Newkome, G. R.; Morris, K. F.; Johnson, Jr., C. S. *Macromolecules* **1994**, *27*, 3464.

34. Malmström, E.; Liu, F.; Boyd, R. H.; Hult, A.; Gedde, U. W., manuscript.

35. Unpublished results, M. Sc. Ronnie Palmgren, Department of Polymer Technology, Royal Institute of Technology, Stockholm, Sweden.

36. (a) Malmström, E.; Liu, F.; Boyd, R. H.; Hult, A.; Gedde, U. W. *Polym. Bull.* **1994**, *32*, 679; (b) Malmström, E.; Liu, F.; Boyd, R. H.; Hult, A.; Gedde, U. W., manuscript.

37. Johansson, M.; Malmström, E.; Hult, A.; Månson, J.-A. E., manuscript.

38. Johansson, M.; Hult, A., submitted to *J. Coat. Techn.*

39. Pettersson, B.; Sörensen, K. *Proceedings of the 21st Waterborne, Higher Solids, & Powder Coating Symposium*, New Orleans, 1994, p. 753.

40. (a) Hult, A.; Malmström, E.; Johansson, M.; Sörensen, K. Sweden Patent 9200564-4; (b) Hult, A.; Malmström, E.; Johansson, M.; Sörensen, K. *WO* 93/17060.

CONSEQUENCES OF THE FRACTAL CHARACTER OF DENDRITIC HIGH-SPIN MACROMOLECULES ON THEIR PHYSICOCHEMICAL PROPERTIES

Nora Ventosa, Daniel Ruiz,
Concepció Rovira, and Jaume Veciana

Advances in Dendritic Macromolecules
Volume 3, pages 27–59
Copyright © 1996 by JAI Press Inc.
All rights of reproduction in any form reserved.
ISBN: 0-7623-0069-8

ABSTRACT

Advantages and drawbacks of dendritic molecular architectures in order to prepare *polymeric organic magnetic materials* are described and thoroughly discussed. Studies with the first generation of two polychlorinated polyradical series of dendrimers revealed that such architectures are compatible with one of the most efficient ways to achieve intramolecular ferromagnetic couplings; i.e., the topological symmetry degeneration approach based on a dynamic spin polarization effect. Nevertheless, the branch-cell hierarchy of these architectures limits in the practice the sizes and effective *S* values of the resulting *super high-spin macromolecules*. The high persistence as well as the particular structural and conformational characteristics of the two studied dendrimer series have permitted us to evaluate the influence of the surface area and fractal dimensionality of the studied high-spin stereoisomers in their ability to interact with the neighboring solvent molecules influencing, thereby, many of their physicochemical properties. These studies have been performed through the use of linear solvation free energy relationships and have demonstrated that the different sizes and shapes are the most relevant molecular parameters in the chromatographic discrimination of the stereoisomers. The same molecular parameters, together with a minor contribution of the polarity, influence on the relative stability and populations of the stereoisomers existing in solution. By contrast, in the case that the cavitational effects are not involved, the unique relevant molecular parameter that discriminate the behavior of the stereoisomers seems to be the polarity. From these studies it can be concluded that the intrinsic fractal character of dendritic macromolecules may play a major role in many of their physicochemical properties.

I. INTRODUCTION

Dendrimers are highly branched three-dimensional macromolecules with a ramified central core unit and with several branch points at each monomer repeating unit leading to structures that have a defined number of generations and functional end groups.[1,2] These materials are unique since they are potentially the most highly branched molecular structures that exist; being accordingly *fractal molecular objects*. Furthermore, since the polymer chains ramify geometrically with successive generations, the concept of densely packed systems arises in which further growth is precluded by steric constraints imposed by the previous generation. The steric congestion in dendrimers is determined by the sizes, shapes, and multiplicities of the central core and of the monomeric branch-junctures as well as by the sizes and shapes of the end groups. As a result of the high degree of branching and a great conformational disorder, large dendrimers adopt a globular shape in which the bonds converge to the central core unit. As size increases, the shape of successive dendrimer generations progresses from open structures to closed spheroids with well-developed internal hollows, ridged and densely packed surfaces, and large changes in their fractal dimensionalities.[1] Thus, an ideal dendritic growth will only occur until a certain limit generation is reached, at which point the steric congestion will prevent further growth. The steric considerations that dictate both the limit generation and the globular shape of dendrimers also affect strongly their ability to interact with other macromolecules and with the neighboring solvent molecules. Consequently, dendrimers have unusual and not yet very well understood physico-chemical properties.[3–5]

Two distinct general routes to dendrimer synthesis have been evolved from the pioneering work of Vögtle and Denkewalter.[6,7] Newkome[8] and Tomalia[1] developed the *divergent route* in which a stepwise, repetitive reaction sequence began at the central core unit of the dendrimer. Later, Fréchet[9] and Miller[10] developed an alternative path, named *convergent route*, in which the synthesis begins at what will become the outer surface of the dendrimer. Nevertheless, both routes proceed by stepwise reiterative reaction sequences in which each addition of monomer (a generation growth step) results in the genesis of what will be a new generation of the dendritic series. This characteristic provides to these materials the capability to be obtained as monodispersed macromolecules with mesoscopic dimensions.

Applications of dendritic macromolecules are presently confined to their use as molecular size standards, or submicron calibrators, due to their monodispersivities and sizes. Nevertheless, several other technological applications for these three-dimensional polymers are currently envisaged.[1] Indeed, materials of this type have been suggested as building blocks of mesoscopic systems for nanotechnology, in which three-dimensional structures may be constructed on a molecular level. Such mesoscopic systems might have broad implications in fields as biophysics, medical diagnosis, and therapy, as well as in microelectronics and materials research.

II. DENDRITIC HIGH-SPIN MACROMOLECULES

In the field of *molecular magnetic materials* there has been an increasing interest in the development of rational and simple routes to bulk ferromagnetic materials and large high-spin clusters.[11] Especially attractive for such purposes are purely organic macromolecules composed of open-shell repeating units having architectures other than linear chains. Examples of such architectures include comb, star, and dendritic polymers as shown in Figure 1.

These molecular architectures, and in particular the dendritic ones, have several advantages from physical and synthetical points of view.[12] First, they are compatible with the most efficient ways to achieve robust intramolecular ferromagnetic interactions between the neighboring open-shell units of a macromolecule. Second, due to their large degree of branching, the establishment of several pathways for the ferromagnetic coupling between open-shell units is easily attained. And, third, these molecular architectures are one of the most practical ways to obtain, by means of a stepwise synthetic procedure with a limited number of steps, macromolecules with a very large degree of polymerization and, therefore, with an extremely large number of ferromagnetically coupled open-shell units.

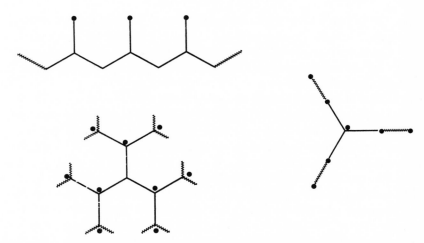

Figure 1. Macromolecular architectures with open-shell repeating units used to build organic/molecular magnetic materials.

In order to build an organic polymer with a bulk magnetic property, like ferro- or superparamagnetism, it is necessary to assemble hundreds or thousands of monomers with a permanent open-shell character (i.e., free radicals, radical ions, carbenes, or nitrenes) in a very large and three-dimensional macromolecule. In addition, the spins of these monomeric-repeating units must be ferromagnetically coupled to each other.[12] Many researchers have investigated several synthetic approaches or strategies to obtain such *super high-spin organic macromolecules*. The most fruitful one developed up to date is the so-called *topological symmetry degeneration* approach.[13] This approach is based on a particular intramolecular magnetic exchange mechanism that operates in so-called non-Kelulé alternant hydrocarbons and is based on a dynamic spin polarization effect.[14] These hydrocarbons are fully conjugated, planar π-systems with appropriate topologies (or connectivities) that produce as many degenerated nonbonded molecular orbitals (NBMO) as the number of open-shell centers in the macromolecule. Since the degeneracy of NBMOs in such systems is only set by their topologies, the question of the spin predilection as well as its robustness can be analyzed solely in terms of the strength of the involved electron exchange repulsions. Electronic repulsions in such π-systems are enlarged when the NBMOs are very much coextensive in space and, thereby, a simple guide to design robust high-spin molecules is to choose the proper topology that produces nondisjoint (i.e., coextensive) NBMOs with as much spatial coextension as possible. In practice, one of the most effective ways to attain such a situation and, therefore, to obtain extremely robust high-spin molecules is to join two or more open-shell centers through positions in which the semi-occupied molecular orbital (SOMO) of each center has a node, as occurs with the central carbon atom of an allyl radical or the *meta* positions of benzylic or triphenymethyl radicals.

This approach to robust high-spin molecules was first used by Itoh et al.[15] some years ago for preparing a series of linear *meta*-substituted polycarbenes **1** ($n = 1-5$; with total spin quantum numbers of $S = 1-5$, respectively) depicted in Figure 2. Later, Rajca et al. prepared several *meta*-substituted polyradicals **2** ($n = 3, 6$, and 9; with $S = 2, 7/2$, and 5, respectively) but using a starbranched topology in this series of high-spin molecules.[16] The same topology was also used successfully by Iwamura et al. for obtaining a series of *meta*-substituted polycarbenes **3** ($n = 3, 6$, and 9; with $S = 3, 6$, and 9, respectively) to which belongs the macromolecule with the largest spin multiplicity, $S = 9$, reported to date.[17]

All linear and starbranched polymers shown in Figure 2 are robust high-spin macromolecules with thermally inaccessible low-spin states. Nevertheless, their molecular sizes and S values are still too small to achieve a long-range ferromagnetic ordering and, accordingly, no polymeric bulk ferromagnet has been prepared thus far.[18] Another disadvantage of such polymers for any practical purposes is their low life expectancies or persistencies. Although they are perfectly stable in isolation (i.e., in diluted frozen solid solutions), they are highly reactive species that do not survive when heated to higher temperatures. In order to overcome these two problems, the use of dendritic architectures was independently proposed.[19,20] Actually, the high branching of these structures would permit us to achieve in few steps and with less synthetic effort very large high-spin organic macromolecules

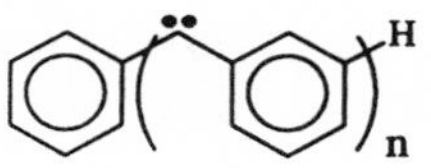

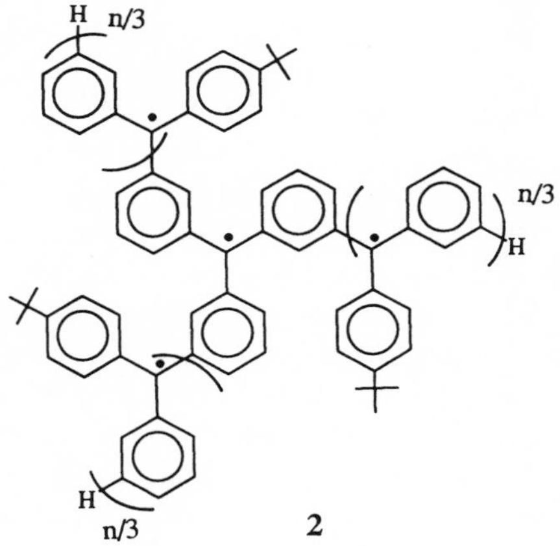

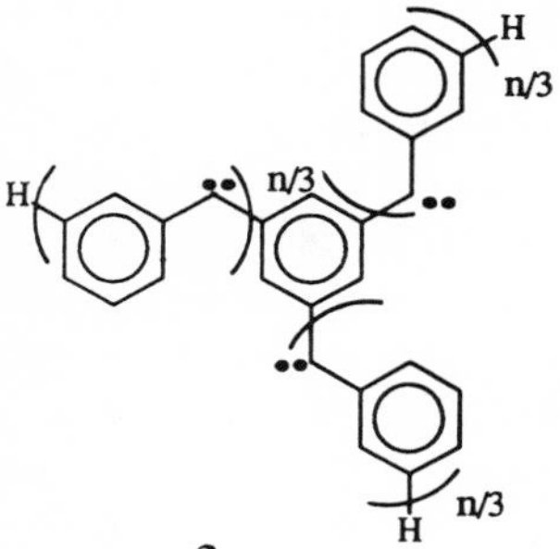

Figure 2. Robust super high-spin organic macromolecules with linear and star-branched topologies.

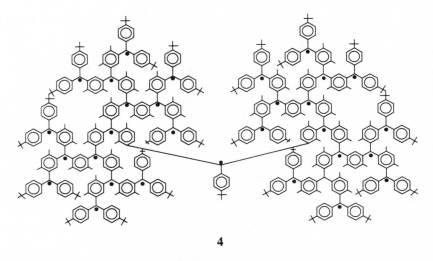

4

Figure 3. Robust super high-spin organic macromolecule with a dendritic architecture based on a three-coordinated Cayley tree.

and at the same time provide enough steric shielding to increase the persistency of their open-shell centers. Attempts to increase the spin multiplicity of organic macromolecules to values larger than $S = 9$, using dendritic architectures based on a three-coordinated Cayley tree (Bethe lattice) like the polyradical **4** shown in Figure 3, were pursued by Rajca et al. However, the resulting super high-spin molecules, one nominally with a $S = 31/2$, showed several defects on their structures that disrupt the theoretically expected ferromagnetic coupling pathways lowering their effective S values to $S \leq 4/2-5/2$.[19]

Our choice was the two series of dendritic polymers **5** and **6,** depicted in Figure 4, which have all their open-shell centers (or trivalent carbon atoms) sterically shielded by an encapsulation with six bulky chlorine atoms in order to increase their life expectancies and thermal and chemical stabilities.[20] Indeed, it is very well known that the monoradical counterpart of both series of polyradicals, the perchlorotriphenyl methyl radical, shows an astonishing thermal and chemical stability for which the term of *inert free radical* was coined.[21] The series of dendrimer polymers **5** and **6** differ in the nature and multiplicity (or branching) of their central core unit, N_c, as well as in their branch-juncture multiplicities, N_b.[1] Thus, series **5** has a hyperbranched topology with $N_c = 3$ and $N_b = 4$, while dendrimer series **6** has a lower level of branching with $N_c = 3$ and $N_b = 2$, and the topology of a three-coordinated Cayley tree.

In order to obtain both series of dendrimers a divergent stepwise synthetic approach was designed.[20a] Each generation growth step of this synthetic approach implies four consecutive reactions: (a-b) two successive alkylation reactions that

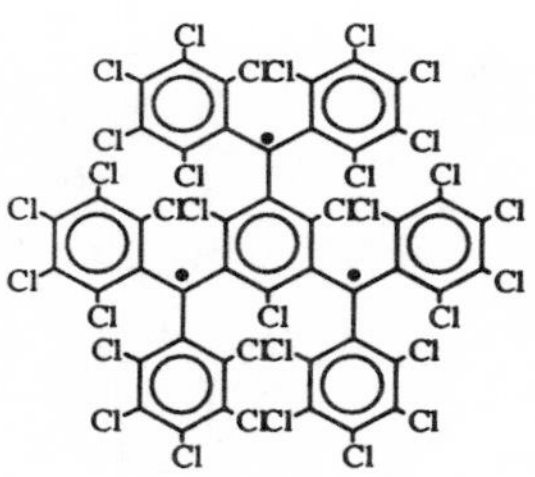

5 (G = 1)
S = 3/2

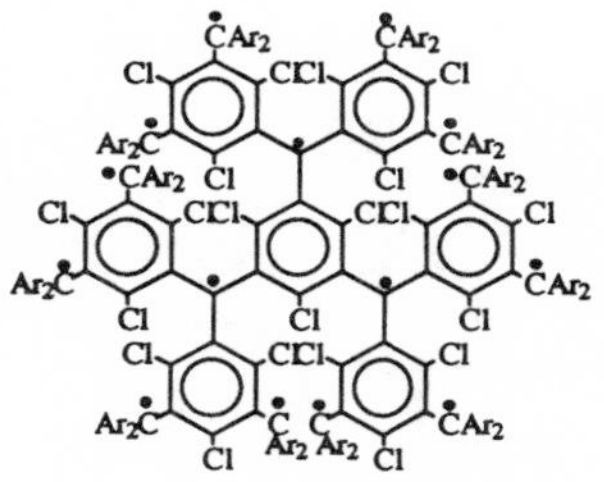

5 (G = 2)
S = 15/2

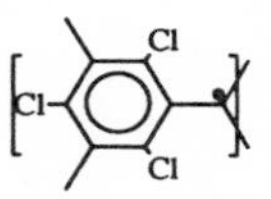

5 (G = n)
S = $(4^n - 1)/2$

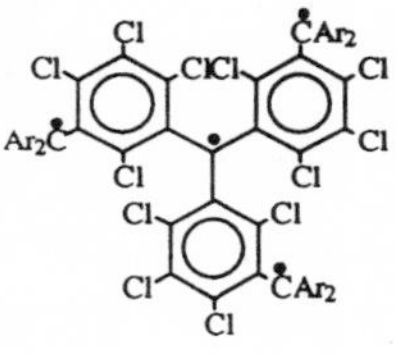

6 (G = 1)
S = 2

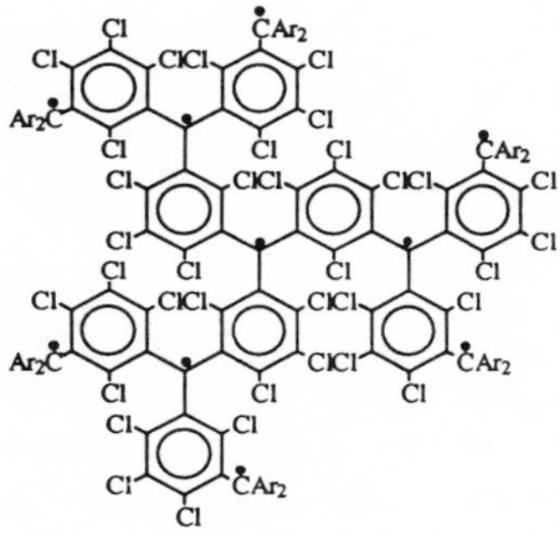

6 (G = 2)
S = 5

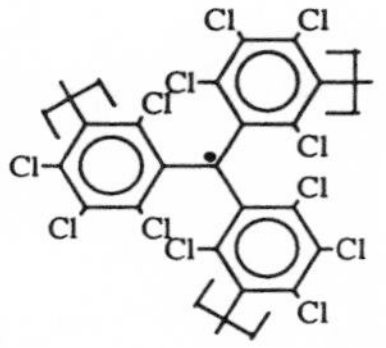

6 (G = n)
S = $3 \cdot 2^{n-1} - 1$

Figure 4. Series of persistent and robust super high-spin organic macromolecules with dendritic nature having a hyperbranched architecture, **5**, and a topology of a three-coordinated Cayley tree, **6**.

yield two different hydrocarbon compounds with several triphenylmethane groups connected through the *meta* positions of their aromatic rings, (c) one acid–base reaction for the generation of carbanions from each of these triphenylmethane groups, and (d) the oxidation of such carbanions to produce the open-shell or radical centers. The reiteration of these four steps would produce the distinct generations (G = 1 – n) of the dendrimer series depicted in Figure 4, where the S values of their ground states, theoretically expected from their *meta*-substitution pattern, are also given.

The first generation (G = 1) of the dendrimer series **5** was obtained without any difficulty using the described synthetic approach above.[22] This open-shell compound exists in two interconvertible pairs of enantiomers differing in the helicity of the three propeller-like conformations adopted by each diphenylmethyl group.

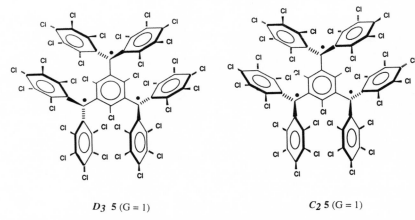

D_3 **5** (G = 1) C_2 **5** (G = 1)

Figure 5. Stereoisomers of the high-spin dendrimer **5** (G = 1) with D_3 and C_2 symmetries which differ in the helicity of their three propeller-like conformations adopted by each diphenylmethyl group.

Such pairs of enantiomers have C_2 and D_3 symmetries, as shown in Figure 5,[23] and are isolable by HPLC chromatography (see Figure 7a) or crystallization as stable solids. Accordingly, with the large steric hindrances and shieldings of the three trivalent carbon atoms, both isomeric forms of the triradical show high energy barriers for their interconversion (in THF at 298 K; $\Delta G^{\neq} = 101 \pm 2$ kJ mol^{-1}) as well as overwhelming persistencies and stabilities.[24] In fact, any significant interconversion is observed in solution at temperatures lower than 273 K. On the other hand, the lifetimes of both isomers are of decades since they do not show any sign of decomposition up to 520 K, even in air. As expected from theoretical considerations based on their *meta*-substitution patterns, the ground states of both isomeric forms are *quartets*.[22]

The second generation of this dendrimer series, **5** (G = 2), was also obtained following the divergent stepwise synthesis previously outlined. As occurred for the dendrimer **4**, the resulting dendritic compound has several structural defects, both in its internal zones and on the external tiers, that interrupt the magnetic interactions among the isolated radical centers and, consequently, originate a drop of the effective S value from the maximum possible value of $S = 15/2$ to $S \leq 1$.[24] This result indicates therefore that the limit generation of the dendrimer series **5** is reached very early—just during the first generation. Sources of the above-mentioned defects are the hyperbranched nature of this molecule with a $N_c = 3$ and $N_b = 4$ that reduces enormously the available surface area per terminal group, as well as the large bulkiness of the branch segments and the terminal groups, i.e., the *m*-trichlorophenylene and pentachlorophenyl groups, respectively. Thus, the large steric congestion developed in the second generation precludes the ideal growth of the hydrocarbon precursor and hinders the entrance of the reagents for subsequent formation of the carbanionic and radical centers, thus explaining the resulting low effective S values.

Concerning the series of dendritic polymers **6**, to date we have only obtained the first generation, i.e., the tetraradical **6** (G = 1). The stereochemical analysis of this open-shell molecule, based on previous considerations given by Mislow et al.[25] and energy minimizations performed by empirical calculations with the Dreiding force-field,[26] predict 12 different enantiomeric pairs for which different relative abundances and slow isomerizations are expected.[27] As occurred for **5** (G = 1), this is due to the highly restricted rotations of the chlorinated aromatic rings, which are caused by the large steric hindrance produced by the bulky *ortho*-chlorine atoms. Figure 6 schematically depicts these 12 stereoisomers which correspond to distinct real energy minima and only differ in the helical senses (*P*, plus, and *M*, minus) of the four propeller-like subunits, as well as in the relative orientations of the three external $Ar_2C^\bullet$ subunits with respect to the central $Ar_3C^\bullet$ subunit.[23]

The presence of these 12 stereoisomers was experimentally confirmed by the partial resolution of 12 peaks in HPLC chromatography using an achiral stationary phase (octadecyl polysiloxane, ODS) and different mobile phase compositions (CH_3CN/THF; 55:45–58:42) and temperatures (288–258 K). Figure 7 shows one of these chromatograms in which 10 partially resolved peaks are clearly distinguished.[28]

The ground state of a solid sample of the tetraradical **6** (G = 1), composed of these 12 stereoisomers, was determined by magnetic susceptibility and magnetization measurements under different conditions.[27] Surprisingly, the effective ground state of **6** (G = 1) was not a *quintet state*, $S = 2$, as theoretically expected from the *meta connectivity of its four open-shell centers, but a triplet* one; $S = 1$. As shown in Figure 8, this result can be rationalized by a Kambe's analysis of a four open-shell center model with a pyramidal topology. The presence in this magnetic model of three antiferromagnetic interactions, $J_1 < 0$, among the external centers, $S_e = 1/2$, that compete with the ferromagnetic interactions, $J_2 > 0$, that take place between the central $S_c = 1/2$ center and the three external centers, will turn upside down the relative energies of the $S = 2$ and $S = 1$ states for ratios of $J_1/J_2 < -0.4$. The presence in **6** (G = 1) of intramolecular ferromagnetic interactions is ascribed to the dynamic spin polarization mechanism due to the *meta*-substitution pattern. The appearance of additional intramolecular antiferromagnetic interactions in this dendrimer could be ascribed to the close contacts existing between some of the atoms and/or rings of the external centers $Ar_2C^\bullet$ which are produced by the large congestion of this molecule. Such close contacts would be responsible for the direct overlap of the SOMOs of the interacting centers, therefore favoring the antiferromagnetic nature of the resulting couplings.[12,27]

The large overcrowding present in the dendrimer series **5** and **6**, as well as in the dendritic polymer **4** obtained by Rajca et al.,[19] show that the use of dendritic architectures to produce super high-spin macromolecules has some practical weak-

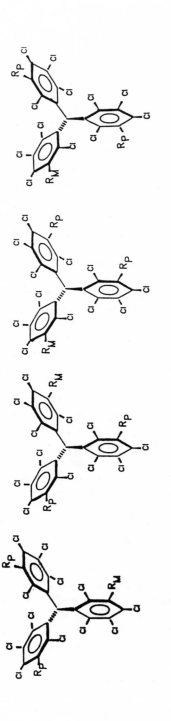

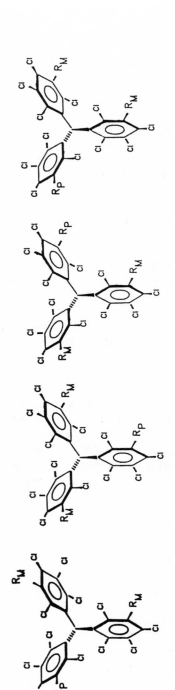

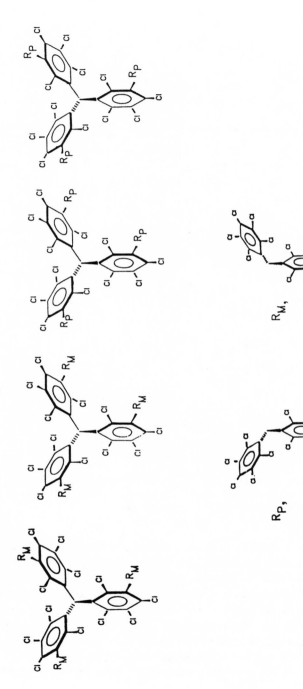

Figure 6. Stereoisomers of dendrimer **6** (G = 1) differing in the helical senses of the four propeller-like subunits and in their relative orientations.

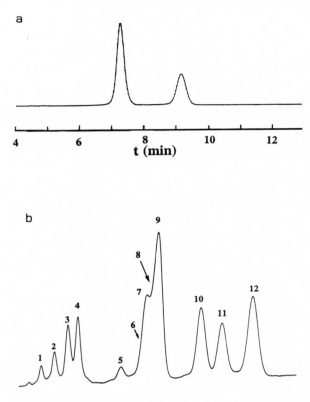

Figure 7. Representative HPLC separations of the stereoisomers of (**a**) dendrimer **5** (G = 1) at 298 K and (**b**) dendrimer **6** (G = 1) at 258 K using a HPLC achiral stationary phase of ODS and ACN/THF (55:45) as mobile phase.

ness. The drawbacks involved in such architectures are always associated with structural and/or magnetic coupling defects that limit the sizes and effective S values of these open-shell macromolecules. Actually, dendritic architectures have a large degree of branch-cell hierarchy. Thus, the interior zones containing cascading tiers of branch cells with radial connectivity to the central core have a higher hierarchy than the exterior zones. Accordingly, any structural defect present in the interior regions of a dendrimer will have larger consequences than in the exterior ones since they will disrupt the magnetic communication among several and wide regions of the macromolecule. In order to overcome these problems two other approaches have

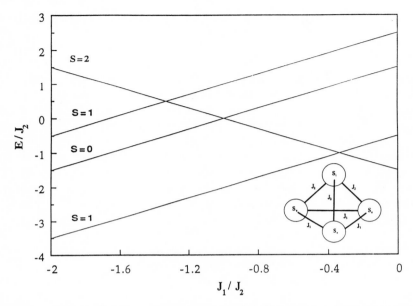

Figure 8. Dependence of the reduced energy, E/J_2, of the ground and excited states on the J_1/J_2 ratio for a four-spin system, with $S_e = S_c = 1/2$ and $J_2 \geq 0$, arranged as in the symmetric tetrameric magnetic model shown.

been recently proposed.[29,30] One of them is based on a dendritic architecture with a particular "closed loop" (or ring) molecular connectivity that maintains the *meta*-substitution pattern and, thereby, each open-shell center has multiple ferromagnetic spin-coupling paths.[29] The other approach requires polymers with π-conjugated skeletons with a proper topology and with a comb-like structure in which each tooth has an open-shell center.[30] Due to the topology of the ferromagnetic coupling paths involved in both approaches, the presence of any defect will not reduce dramatically the total spin values since it will have only a local effect. In spite of this remarkable advantage, the largest spin values achieved up to date with both approaches are still low: $S = 4$ and $S \leq 3/2$, respectively. In both approaches the origin of these low S values seem to be associated with synthetic difficulties rather than to other causes.

The above-mentioned results and drawbacks made futile any effort to grow larger generations of the dendrimer series **5** and **6** in order to get super high-spin macromolecules. Nevertheless, the outstanding structural characteristics and physicochemical properties of the first generations of both dendrimer series, **5** (G = 1) and **6** (G = 1), make both modest dendrimers worthy of detailed physicochemical studies.

III. FRACTAL CHARACTER OF DENDRITIC HIGH-SPIN MOLECULES

It is well known that dendrimers exist in an enormous number of conformations with similar relative abundances due to their high flexibilities, large number of degrees of freedom, and comparable energies.[1] Therefore in the condensed phase each dendrimer is expected to have a conformation different from the next, so there are no possibilities of long-range orderings with such kind of materials. Consequently, both in the condensed phase and in solution, the different generations of dendrimers consist of extremely complex, equilibrated mixtures of all of their conformers resulting in collections of fractal molecular objects. On the other hand, there are many studies revealing that the interactions of a macromolecule with its surroundings and, thereby, many of its properties (e.g.; solubility, hydrophobicity, and thermochemical activity, etc.) are modulated first and foremost by its molecular size and shape.[31,32] Therefore, the interactions with solvent molecules of dendritic macromolecules will be influenced by their intrinsic fractal character hindering enormously the study of their physicochemical properties.

High-spin dendrimers **5** (G = 1) and **6** (G = 1) are exceptional in the sense that they exist in several isolable and stable stereoisomeric forms (2 and 12 respectively) that show striking differences in their overall sizes and shapes as well as in their external surfaces. Furthermore, these stereoisomeric forms are interconverting at higher temperatures, but these processes can be stopped at will just by lowering the temperature leading then to equilibrated mixtures of all of their isomers.[22] Since the structural features of the stereoisomers of **5** (G = 1) and **6** (G = 1) are well-defined for each stereoisomer and are responsible, together with their own electronic characteristics, for their different physicochemical properties, such dendrimers are ideal models for studying the influence of certain molecular structural parameters, like the surface area, volume, and fractal dimensionality in the ability of dendritic macromolecules to interact with neighboring solvent molecules. Consequently, dendrimers **5** (G = 1) and **6** (G = 1) provide a valuable opportunity to study the influence of their fractal character on their properties. As physicochemical properties for this study we chose the differential chromatographic retentions of the distinct diastereoisomers of **5** (G = 1) and **6** (G = 1) as well as the dependence of the equilibrium constant and of the tumbling rates of the two isomers of **5** (G = 1) on the nature of the surrounding medium. Before giving an overview of the first of such studies, a brief description of the most relevant structural and electronic characteristics of the stereoisomers of **5** (G = 1) and **6** (G = 1) is required.

A. Structural Characteristics of Dendrimers

In order to give a description of the overall size and shape as well as the outside surface characteristics of a dendrimer, some quantitative parameters are required. A general idea about the overall shape of a dendrimer can be inferred by the aspect

Table 1. Structural and Electronic Characteristics of Stereoisomers of Dendrimer **5** (G = 1) Obtained by AM1 Semiempirical Calculations

Isomer	ΔH[a]	ΔS[b]	I_z/I_x (I_z/I_y)	μ[c]
C_2	371.28	$-R\ln 2$	1.38 (1.32)	0.0527
D_3	370.34	$-R\ln 6$	1.40 (1.40)	0.00

Notes: [a] In kJ mol^{-1}.

[b] Estimated by symmetry considerations, in J K^{-1} mol^{-1}, see Ref. 34.

[c] Dipolar moments, in Debyes.

ratio of the largest to smallest principal inertia moments (I_z/I_x and I_z/I_y). Aspect ratios close to 1.0 are typical of spherical shapes, while larger values indicate irregular shapes. For most of dendrimer series the aspect ratio generally decreases from a large value (e.g., I_z/I_x = 4–5) for the generation G = 0 to values close to 1 for higher generations. Consequently, the change of this parameter illustrates that the early generations are somewhat amorphous in shape and the later ones are clearly spherical, especially if they are close to the limit generation.[1]

Table 1 shows the aspect ratio values for the geometries of the stereoisomers of dendrimer **5** (G = 1) with D_3 and C_2 symmetries, obtained by semiempirical AM1 optimizations.[33] Other molecular characteristics, like their molar enthalpies of formation, entropies, and dipolar moments, which are relevant for a complete characterization of these isomers, are also given in Table 1. The aspect ratios of both isomers are very low, close to 1.3, indicating therefore spheroidal shapes in both cases. This fact is in complete agreement with the results previously described, suggesting that the first generations of both dendrimer series are close to the limit ones.

A more precise description of the overall size and shape as well as of the outside surface characteristics of a dendrimer can be made by using the concept of *solvent-accessible surface* (SAS).[35,36] The SAS is obtained by rolling a sphere of radius r—the probe radius—around the van der Waals surface of the molecule, where r represents the effective radius of the solvent. The SAS is then composed of the locus over which the probe sphere rolls. The surface area defined by the calculated SAS, A_{SAS}, and the volume contained within this surface, V_{SAS}, can be considered respectively as the effective accessible area and the excluded volume that the molecule will exhibit to a solvent with an effective radius of r. The surface area is a continuous function of the probe radius, $A_{SAS} = f(r)$, that can be calculated by different algorithms and programs,[36,37] and is extremely useful for studying nonspecific solute/solvent interactions. Another structural feature relevant to such interactions is the texture (roughness) of the solute surface.[38,39] The degree of irregularity of a molecular surface may be described by the *fractal dimension D*,[40] where $2 \leq D \leq 3$. As a surface becomes more irregular, the fractal dimension

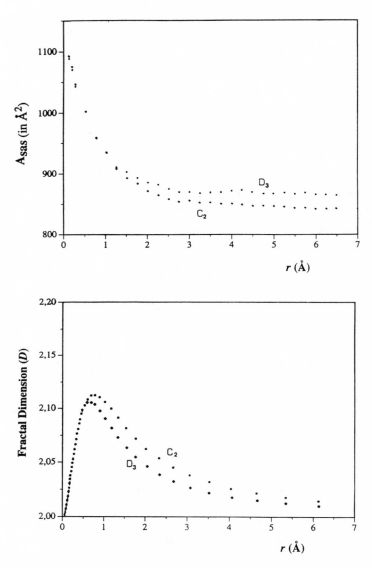

Figure 9. (a) Dependence of molecular surface area on probe radii for D_3 and C_2 isomers of dendrimer **5** (G = 1). (b) Dependence of fractal dimension, D, on probe radii for the same isomers. The derivative in Eq. 3 was numerically approximated from the data illustrated in (a).

increases from the value $D = 2$, for a smooth surface, to $D \leq 3$. Accordingly with Eq. 1, the value of $(2 - D)$ may be obtained from the slope of the plot of $\log[A_{SAS}]$ against $\log(r)$, used to define the SAS.

$$2 - D = d\log[A_{SAS}(r)]/d\log(r) \tag{1}$$

Such a relationship is illustrated in Figure 9 for the AM1 optimized geometries of the stereoisomers of dendrimer **5** (G = 1) with D_3 and C_2 symmetries.[33] The slopes of both $A_{SAS}(r)$ function in the two plots approach zero (corresponding to $D = 2$) in the limit of both small and large probe sizes. Small probes (for $r << 0.1\text{Å}$) predominantly interact with the smooth van der Waals spheres describing the atomic framework of the stereoisomers, whereas large probes (for $r > 7\text{Å}$) are sensitive only to the overall shapes of the two isomers which are quite spherical and show distinct effective diameters. The effective diameter, calculated in such a way, for the D_3 isomer is somewhat larger than for the C_2 isomer as shown in Figure 9a. For probes with radii of 1.0 to 3.5 Å, however, the A_{SAS} and D values of both isomers are quite different being always larger the fractal dimension of the C_2 isomer; this result is in accordance with its larger roughness, lesser molecular surface area, and smaller effective diameter. Since this probe size range corresponds to the approximate sizes of organic solvent molecules and of the side chains of most of the chromatographic stationary phases, such probes should give different intermolecular interactions with both isomers of dendrimer **5** (G = 1).

The same situation is expected for stereoisomers of the dendrimer **6** (G = 1), as revealed by a careful inspection of Figure 10 in which the dependence of the fractal dimension of the 12 isomers on probe radii is depicted. Due to the large number of atoms of this open-shell dendrimer, a low level theoretical method, the Dreiding force-field,[26] was used for the optimizations of the geometries of these 12 isomers.[27]

B. Solvent Effects and Linear Solvation Free Energy Relationships

For a complete quantitative description of the solvent effects on the properties of the distinct diastereoisomers of dendrimers **5** (G = 1) and **6** (G = 1), a multiparameter treatment was used. The reason for using such a treatment is the observation that solute/solvent interactions, responsible for the solvent influence on a given process—such as equilibria, interconversion rates, spectroscopic absorptions, etc.—are caused by a multitude of nonspecific (ion/dipole, dipole/dipole, dipole/induced dipole, instantaneous dipole/induced dipole) and specific (hydrogen bonding, electron pair donor/acceptor, and charge transfer interactions) intermolecular forces between the solute and solvent molecules. It is then possible to develop individual empirical parameters for each of these distinct and independent interaction mechanisms and combine them into a multiparameter equation such as Eq. 2,[41]

$$XYZ = (XYZ)_0 + aA + bB + cC + \ldots \ldots \tag{2}$$

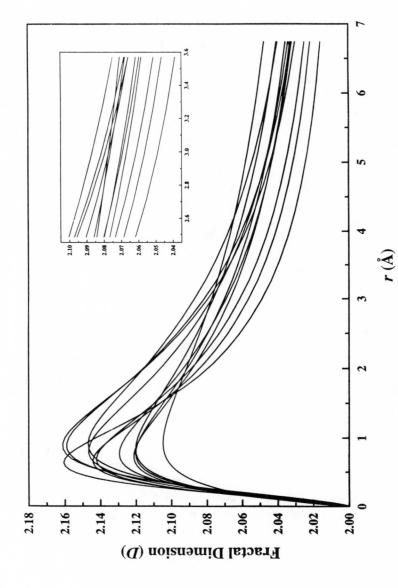

Figure 10. Dependence of fractal dimension, D, on probe radii for the 12 isomers of dendrimer **6** (G = 1). The curves were obtained from the corresponding computed molecular surface areas, not depicted. The insert shows an enlargement of this dependence for a probe size range corresponding to the approximate sizes of organic molecules.

44

where the regression coefficients, $a, b, c, \ldots$, describe the sensitivity of the solute property XYZ to the different solute/solvent interaction mechanisms, whereas the independent variables, $A, B, C, \ldots$, measure the capability of each solvent to interact through a given mechanism. The aforementioned independent terms must be orthogonal and the assumption is made that there is a linear free energy relationship between each term and the specified solvent effect. Finally, the independent term $(XYZ)_o$ is the solute property in the absence of any solute/solvent interaction and, therefore, this term is a good approximation to the expected value of the solute property in the vacuum.

One of the most ambitious, and very successful, quantitative treatments of solvent effects by means of a multiparameter equation, such as Eq. 1, is that developed by Kamlet and Taft and called *linear solvation free energy relationship* (LSER).[42] Such a quantitative treatment assumes that the attractive (or exoergic) interactions can be correctly categorized by a nonspecific interaction term that describes the effect of dipolarity, and other specific terms that characterize, among others, the effect of hydrogen-bond donor and acceptor interactions on the solvent properties. Finally, to complete the description of the solvent effect, a nonspecific endoergic term that describes the formation of a cavity in the solvent with the proper size and shape to accommodate each solute molecule must also be included in this model. The energy, which is required to produce a cavity in the bulk of the solvent to include a solute molecule, is proportional to the surface area of the solute; the curvature or texture of this molecular surface determines the surface work required, which is an important component of the energy associated with this process.[32]

Usually, linear solvation free energy relationships adopt the form of Eq. 3,[41]

$$XYZ = (XYZ)_o + s(\pi^* + d\delta) + a\alpha + b\beta + m\Omega \tag{3}$$

where $(XYZ)_o$, s, d, a, b, and m are solvent-independent coefficients characteristic of the process and solute under study and indicative of its susceptibility to the solvent properties. Thus, the relative values of the coefficients s, a, b, and m describe the relative importance of dipolar and hydrogen-bonding interactions and of the cavity formation on the studied property XYZ. The solvent attributes in Eq. 3 are given by the UV–Vis spectroscopically derived solvatochromic parameters π^*, δ, α, and β, and by the cavitational parameter Ω which take into account the different solute/solvent interaction mechanisms. Thus, the solvatochromic parameter π^* measures the exoergic effects of dipole/dipole and dipole/induced dipole interactions between the solute and solvent molecules, while the parameter δ is just a factor that corrects the parameter π^* for certain families of solvents. Therefore, both parameters measure the ability of a solvent to stabilize a neighboring charge or dipole by virtue of nonspecific dielectric interactions. The solvatochromic parameter α is a quantitative, empirical measure of the ability of a bulk solvent to act as a hydrogen-bond donor toward a solute. By contrast, the empirical parameter β measures quantitatively the ability to act as a hydrogen-bond acceptor or electron-pair donor toward a solute. That is, it measures the capability to behave as a hydrogen-bond donor or a solvent-to-solute coordinative bond, respectively. In addition to the solvatochromic parameters, which represent the exoergic solute/solvent interactions, Eq. 3 includes an extra term, called cavity term in which the parameter Ω represents the physical quantity named cohesive pressure (or cohesive energy density) of the solvent. This quantity is the square of the Hilderbrand's solubility parameter, δ_H, which is given by $\delta_H = (\Delta H^0 - RT/V_m)^{1/2}$, where ΔH^0 is the molar standard enthalpy of vaporization of the solvent to a gas of zero pressure, and V_m is the molar volume of the solvent.[43]

As already mentioned, the cavity term corresponds to the endoergic process of separating the solvent molecules to provide a suitably sized and shaped enclosure for the solute, and measures the work required for such a purpose. This term is related to the tightness or structuredness of solvents as caused by intermolecular solvent/solvent interactions. The association of solvent molecules in the liquid state in order to accommodate the solute molecules can be quantified by means of the surface area and texture of the solute that are related with the m coefficient and by the cohesive pressure of the solvent given by Ω.

Most of the reported LSERs are simpler than that indicated by Eq. 3. For example, if the solvent-induced change of the property does not involve the creation of a cavity or a change in the shape of the cavity between the different states of the studied process (e.g., the limit states of an equilibrium, the initial and transition states of a kinetic process, or the ground and excited states of a spectroscopic absorption) the cavity term drops out. This is the situation that occurs for most of the spectroscopic measurements due to the Franck–Condon principle. Alternatively, if not hydrogen-bond donor solvents are considered, the exoergic term of Eq. 3 containing the α-parameter also drops out. In this way, depending on the nature of the studied solute and solvents, the four terms of Eq. 3 can be reduced to only three, two, or, even, one term. Equation 3 applies to the influence of *different solvents* on the property *XYZ* of a *single solute* (e.g. an equilibrium constant, a process rate, a spectroscopic absorption coefficient, etc.). Conversely, the same equation can also be used to correlate the property *XYZ* of a set of *different solutes* in a *single solvent* (e.g. solubility, partition coefficient, spectroscopic absorptivity, etc.), providing the meaning of the coefficients and parameters of Eq. 3 is changed. In this particular case the parameters π^*, δ, α, β, and Ω represent solute parameters, not solvent parameters, while $(XYZ)_o$, s, d, a, b, and m are solute independent coefficients, characteristic of the process under study and indicative of its susceptibility to the nature of the solute.

The LSER methodology has been used to unravel, recognize, and rank the individual solute/solvent interactions that determine the solvent effects of an extremely large number of examples.[42,44] Nevertheless, no examples of such a kind of studies with dendritic macromolecules have been reported to our knowledge.

C. Physicochemical Properties of Dendrimers

As already mentioned, we chose three different physicochemical properties for studying the influence of the surface area and fractal dimension in the ability of dendritic macromolecules to interact with neighboring solvent molecules. These properties are: (a) the differential chromatographic retention of the diastereoisomers of **5** (G = 1) and **6** (G = 1), (b) the dependence on the nature of solvents of the equilibrium constant between the two diastereoisomers of **5** (G = 1), and (c) the tumbling process occurring in solution of the two isomers of **5** (G = 1), as observed by electron spin resonance (ESR) spectroscopy. The most relevant results and conclusions obtained with these three different studies are summarized as follows.

Differential Chromatographic Retentions of Dendritic Molecules

Our initial attempts to separate the C_2 and D_3 isomers of **5** (G = 1) used octadecyl polysiloxane (ODS) as high performance liquid chromatography (HPLC) stationary phase and mixtures of acetonitrile/H_2O or methanol/H_2O as mobile phases. Under these classical reverse-phase conditions, the resulting efficiencies were extremely poor because of the low solubility of **5** (G = 1) in both mobile phases. By contrast, mixed mobile phases which contained acetonitrile (ACN) with some percentages of a cosolvent such as tetrahydrofuran (THF) substantially improved

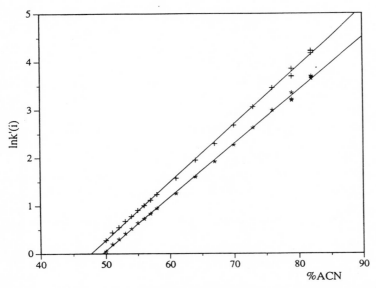

Figure 11. Variation of capacity factors of the D_3 (+) and C_2 (*) isomers of dendrimer **5** (G = 1) with percent ACN in THF as a mobile phase using ODS as stationary phase at K.

the efficiency, producing considerable retentions and a complete separation of both isomers. Figure 7a is a sample chromatogram using such conditions in which two peaks corresponding to the isomers with C_2 and D_3 symmetries are shown. Figure 11 illustrates how the retention of these two isomers is affected by the addition of increasing amounts of ACN to THF. The logarithm of the capacity factors, ln k', of both isomers increased with the percentage of ACN added to THF, in that the changes for the D_3 isomer are slightly higher. Nevertheless, with percentages of ACN below 50% and down to pure THF, both diastereoisomers were unretained.

Temperature had also a noticeable influence on the capacity factors and the separation factor, α, of the two isomers of **5** (G = 1), as illustrated in the series of chromatograms depicted in Figure 12. It appears that with an increase in temperature, the retention of the more strongly retained isomer, the D_3 one, decreases at a greater rate than the retention of the C_2 isomer. Thus, the ln α drops as the temperature decreases in the range of 280–320K.

This temperature effect is the usual one for a "regular" or "enthalpy–entropy compensated" chromatographic separation, suggesting that the retention of each

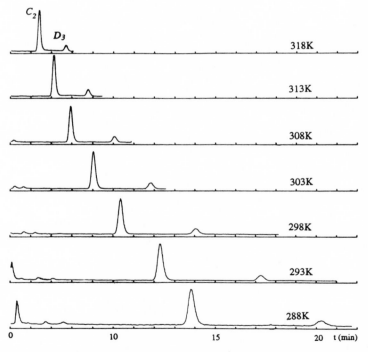

Figure 12. Chromatographic resolutions at different temperatures of the D_3 and C_2 isomers of dendrimer **5** (G = 1) performed with CH_3CN/THF (60:40) as mobile phase and ODS as the stationary phase.

isomer in the stationary phase is controlled by a single operative sorption mechanism and that this mechanism is the same for both isomers.[45,46] Under such circumstances, the capacity factor of each isomer, i, must be related with the changes of standard molar enthalpies, $\Delta\Delta H_i^0$, and entropies, $\Delta\Delta S_i^0$, involved during the transfer of the isomer from the mobile phase to the stationary phase by means of Eq. 4,

$$\ln k_i' = -\Delta\Delta H_i^0/RT + (\ln \phi + \Delta\Delta S_i^0/R) \tag{4}$$

where ϕ is the ratio between the stationary volumes, V_s, and the mobile, V_m, phases of the chromatographic system, i.e., $\phi = V_s/V_m$. Moreover, there must be a linear dependence of $\Delta\Delta H_i^0$ on $\Delta\Delta S_i^0$, such as $\Delta\Delta H_i^0/\Delta\Delta S_i^0 = T_\beta$, in which the so-called isoequilibrium temperature, T_β,[47] is independent of the isomer.

In accordance with Eq. 4, the plots of $\ln k'$ versus $1/T$ for the two isomers of dendrimer **5** (G = 1) appear to be linear in the temperature range studied (r^2 = 0.998–0.999), confirming the previous suggestion that a single sorption mechanism is operative for both isomers. These plots also permit us to calculate $\Delta\Delta H_i^0$ and $\Delta\Delta S_i^0$ values for both isomers, which are −29.5(2) kJmol^{-1} and −85(1) JK^{-1} mol^{-1}, respectively, for the D_3 isomer and −25.2(2) kJmol^{-1} and −74(1) JK^{-1} mol^{-1} for the C_2 one.[48] Interestingly, the two isomers show the very same isoequilibrium temperature, T_β = 344(7) K, indicating that both have the same retention mechanism.

To better understand this common retention mechanism, as well as the solute–solvent interactions involved during the separation, a linear solvent-free energy relationship methodology was used. In fact, Sadek et al. demonstrated that LSERs can be quite useful in describing the partitioning of solutes in HPLC.[49] For such a purpose, generalized LSERs that describe the transfer of a solute from the mobile to stationary phase are required. In chromatographic separations performed at a constant temperature and changing the composition of the mobile phase, the generalized LSER adopts the form of Eq. 5 for a series of solutes,[24]

$$\ln k_i' = e_i' + s_i'\pi^* + a_i'\alpha + b_i'\beta + m_i'\Omega \tag{5}$$

where the e_i', s_i', a_i', b_i', and m_i' are coefficients characteristic of each solute i and independent of the nature of the stationary and mobile phases. Such coefficients must have negative (or positive) signs accordingly with the exoergic (or endoergic) nature of each term. On the other hand, the variables π^*, α, β, and Ω are the solvatochromic and cavitational parameters of the mobile phase for each composition assayed. Finally, the independent e_i' term depends on the nature of both the solute and the mobile and stationary phases. For chromatographic separations in which the composition of the mobile phase is maintained constant but the temperature is changed, the LSER has a distinct form than of Eq. 5 since the temperature changes affect simultaneously to both chromatographic phases. In this case the generalized LSER is as follows,[24]

$$\ln k_i' = e_i^{s/m} + s_i\pi^*(T)^{s/m} + a_i\alpha(T)^{s/m} + b_i\beta(T)^{s/m} + m_i\Omega(T)^{s/m} \qquad (6)$$

in which $\pi^*(T)^{s/m}$, $\alpha(T)^{s/m}$, $\beta(T)^{s/m}$, and $\Omega(T)^{s/m}$ are the differences between the parameters corresponding to the mobile and stationary phases at a given temperature T, i.e., $\Omega(T)^{s/m} = \Omega(T)^s - \Omega(T)^m$. Coefficients $e_i^{s/m}$, s_i, a_i, b_i, and m_i have their usual meaning and their signs must be opposed to those of Eq. 5, which in this case is positive (or negative) accordingly with the exoergic (or endoergic) nature of each term. Consequently, multivariable linear regressions of the experimental data to Eqs. 5 and 6 will provide information about the interactions that are established between the solutes and the two chromatographic phases.

The π^*, α, β, and Ω solvent strength parameters at 283 K for ACN/THF mixtures, with compositions in the range of 50:50–85:15, as well as the $\pi^*(T)^{s/m}$, $\alpha(T)^{s/m}$, $\beta(T)^{s/m}$, and $\Omega(T)^{s/m}$ parameters of a 60:40 ACN/THF mixture in the temperature range of 288–318 K, were experimentally determined by the usual procedures.[41–44,50] When the capacity data of Figures 5 and 6 were fit by a multiple regression procedure to these experimental solvent strength parameters, models that included the dipolarity and the two hydrogen-bonding terms (i.e., complete Eqs. 5 and 6) were inadequate. They did not pass the usual statistical tests even at low confidence levels. Only models, which included exclusively the independent and the cavity formation terms, fit the data well. These models were valid at the 99.95% significance level ($F = 25,000$; $p \leq 0.0005$) test. Table 2 lists the resulting regression coefficients of such fits. The signs of the resulting m_i and m_i' coefficients are negative and positive, respectively, as expected from the endoergic nature of the cavitational term and the theoretical grounds of Eqs. 5 and 6. This indicates, therefore, that any change of temperature or composition of the mobile phase that causes an increase of the energy necessary to form a cavity in the mobile phase (Ω parameter) produces an increase of the retention of both isomers since the transfer to the stationary phase is energetically favored. The relative magnitudes of m_i and m_i' regression coefficients for the isomers D_3 and C_2 are also very interesting. For the two performed chromatographic experiments, the absolute

Table 2. LSER Models[a] Describing the Chromatographic Retention of the Isomers of Dendrimer **5** (G = 1)

	Coefficients							
	e_i'	m_i'	r[b]	n[c]	$e_i^{s/m}$	m_i	r[b]	n[c]
C_2	−24.2	+29.9	0.999	10	−8.4	−14.8	0.999	8
D_3	−26.1	+32.4	0.999	10	−9.6	−17.2	0.999	8

Notes: [a]Models: $\ln k' = e_i' + m_i'\Omega$ and $\ln k' = e_i^{s/m} + m_i\Omega(T)^{s/m}$; values of the cavitational parameters, Ω, were divided by 100 in both linear regressions.
[b]Correlation coefficients.
[c]Number of data points.

values of coefficients for the D_3 isomer are always larger than for the C_2 one, indicating that the first isomer is more sensitive to the cavity formation. This result is in agreement with the larger molecular surface area of the D_3 isomer due to its more openness structure or lesser rugosity.

Therefore, such results clearly demonstrate that the cavity formation process is the mechanism that controls the retention of the isomers of dendrimer 5 (G = 1), and their distinct sizes and shapes are the molecular characteristics responsible for their discrimination.

Comparable conclusions were also achieved from the chromatographic separations of the 12 stereoisomers of dendrimer 6 (G = 1) performed either at a constant temperature with different ACN/THF mixtures or at distinct temperatures but using a fixed mobile phase composition. Indeed, the capacity data corresponding to these isomers fit nicely to LSERs that include only the independent and cavitational terms, as occurs for dendrimer 5 (G = 1). There is also a linear regression between the resulting m_i and m_i' regression coefficients and the calculated molecular surface areas of the resolved isomers, indicating that those with larger surface areas are more sensitive to changes on the solvent strength of the mobile phases. Moreover, this linear regression permits the assignment of each isomer to each observed chromatographic peak. Even more interesting is the analysis of the thermodynamic parameters corresponding to the transfer of the isomers of 6 (G = 1) from the mobile phase to the stationary phase obtained from the plots of ln k' versus $1/T$ for each isomer. As shown in Figure 13, the resulting changes of the standard enthalpies are linearly related with those of the entropies for all the resolved isomers giving an isoequilibrium temperature of 358 ± 22 K which is similar to that of the isomers of

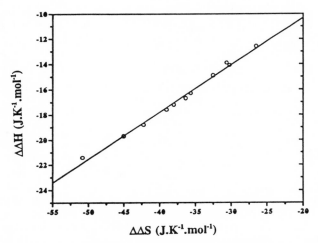

Figure 13. Dependence of $\Delta\Delta H^0$ on $\Delta\Delta S^0$ for the different isomers of the dendritic high-spin molecule 6 (G = 1) in the studied chromatographic system.

dendrimer **5** (G = 1). Hence, discriminations in this family of dendritic high-spin molecules with this chromatographic system are again exclusively due to the differences in their shapes and sizes.

Conformational Populations in Dendritic Molecules

The equilibrium constant for the isomerization process between the stereoisomers of dendrimer **5** (G = 1), $K = [D_3]/[C_2]$ in $C_2 \rightleftharpoons D_3$, shows a dependence with both the temperature and nature of the solvent.[24] Thus, for example, the equilibrium constant in THF at 293 K is 0.40 while at 330 K it diminishes slightly down to 0.34. Changes of this constant are however larger when the nature of the solvent is modified; as exemplified by the value of 0.10 obtained in benzonitrile at 330 K. Both dependencies suggest therefore that the molar standard enthalpies of formation of both isomers are somewhat distinct and that they can be modified by changing the surrounding medium.[51] As a consequence, both isomers must interact with different strengths with the neighboring solvent molecules.

In order to recognize and rank the individual solute/solvent interactions responsible of such solvent effects, a LSER treatment was also used. For such a purpose we choose as representative solvents 17 distinct organic solvents which have a broad range of solvent strength parameters and in which **5** (G = 1) has a significant solubility.[24,52] The π^*, α, β, and Ω solvent strength parameters as well as the isomeric population of the stereoisomers of this dendrimer were determined at 330 K for each one of these solvents. The logarithm of the resulting equilibrium constants was then fit to the model given by Eq. 3 in which the term that includes the β-parameter has been removed.[53] Table 3 lists the regression coefficients and standard errors achieved with this model which is valid at the 99.95% significance level ($F = 19.4$; $p \leq 0.0005$). The signs of these coefficients are as theoretically expected from the endo- and exoergic nature of the respective terms and from the structural and electronic characteristics of both isomers. Indeed, the signs of s and

Table 3. LSER Model[a] Describing the $C_2 \rightleftharpoons D_3$ Isomerization Process at 330 K for **5** (G = 1) on Different Solvents

Proc.	Regression Coefficients[b]						
	e	s	d'[c]	a	m[d]	r[e]	n[f]
$C_2 \rightleftharpoons D_3$	−0.8(2)	−0.3(2)	−0.3(2)	0.7(4)	−0.7(3)	0.930	17

Notes: [a]Model: $\ln K = e + s\pi^* + d'\delta + a\alpha + m\Omega$.

[b]Numbers in parentheses are the standard errors of the corresponding regression coefficients.

[c]The coefficient d' is related with those of equation 2 by $d' = sd$.

[d]Cavitational parameter Ω was divided by 100 to keep its value in the same range as the other strength parameters.

[e]Correlation coefficients.

[f]Number of studied solvents or data points.

d' ($d' = sd$) coefficients are negative, indicating that an increase of the solvent polarity shifts the equilibrium towards C_2 in accordance with its higher dipolar moment (see Table 1). The sign of m is also negative showing that an increase of the solvent cavitational parameter again displaces the equilibrium in the same direction—a result that is in agreement with the smaller molecular surface area of the C_2 isomer (see Figure 9a). On the other hand, the a-coefficient is positive, indicating that a rise of the ability of the solvent to act as a hydrogen-bond donor shifts the equilibrium towards the D_3 isomer. This displacement also agrees with the fact that the chlorine atoms of D_3, which are able to act as hydrogen-bond acceptors, are more accessible to the solvent molecules due to its more structural openness character. The relative magnitudes of these regression coefficients are very important since they permit ranking the individual solute/solvent interactions that are operative in this equilibrium process. The regression coefficient corresponding to cavity formation is the largest one, followed in importance by that associated with the dipolarity, and finally, by the one associated with hydrogen bonds. Finally it is also instructive to analyze the independent or intercept term, e, of the model used to describe the $C_2 \rightleftharpoons D_3$ process on different solvents (see Table 3). As previously mentioned, this term should represent the logarithm of the equilibrium constant at 330 K in the absence of any dipolar, hydrogen bonding, and cavitational interactions. In other words, this term could be treated as a measure of the equilibrium constant at 330 K in the absence of a solvent or its value in vacuum, i.e., $e = \ln K_{330}^0$. Since this constant is related with the differences in the molar standard enthalpies ($\Delta\Delta H^0$) and entropies ($\Delta\Delta S^0$) of formation for both isomers by Eq. 7, it is possible to estimate $\Delta\Delta H^0$ taking as $\Delta\Delta S^0$ the contributions to entropy caused by the molecular rotational degrees of freedom.[34]

$$\Delta\Delta H^0 = -RT\ln K_T^0 + T\Delta\Delta S^0 \tag{7}$$

The relative stability of both isomers estimated in such a way is $\Delta\Delta H^0 = \Delta H^0(D_3) - H^0(C_2) = -1.3 \pm 0.5$ kJmol^{-1}, which agrees with that theoretically calculated by a semiempirical AM1 method, $\Delta\Delta H^0 = -0.9$ kJmol^{-1} (see Table 1).

As a main conclusion of this LSER study one might state that the relative stability of the stereoisomers of **5** (G = 1) in solution is controlled first by the differences in their shapes and sizes and, second, by their distinct polarities. Therefore, the overall size and shape are again one of the most important characteristics of this dendrimer since they play an major role in the relative population of its conformers.

Tumbling Processes in Dendritic Molecules

In order to study a physicochemical property of a dendritic molecule which does not involve either the creation of a cavity in the solvent or the change in its shape, we chose as representative a property that could be studied by a spectroscopic technique such as the molecular tumbling phenomena. Such phenomena occur both in liquid solutions as in solid or viscous amorphous matrices and involve the reorientational motions of the molecules inside the solvent cavities in which they are enclosed. Therefore, the study of the kinetics of such dynamic phenomena will provide information about the solute/solvent interactions that control the orientational–motional correlation times, τ, as well as the associated energy barriers of such a tumbling process.

High-spin dendrimers, like **5** (G = 1), are ideal probes for such a study since they are ESR active and this spectroscopy has the same time scale as such a dynamic process. In fact, the anisotropic components of some ESR parameters, such as those responsible for the fine structures of a high-spin molecule, are averaged out under molecular tumbling leading to noticeably lineshape effects.

Figure 14 shows the first-derivative ESR spectra observed in the temperature range from 100 to 275 K for a benzene solution of the D_3 isomer of dendrimer **5** (G = 1). Changes on these spectra are typical of the lineshape effects produced by reorientational motions of the quartets.[54] Thus, in the *rigid limit region*, that is at temperatures below 130 K, the spectra correspond to those expected for an ensemble of randomly oriented rigid quartets with axial symmetries. They show two extreme turning points or steps, at B_l^e and B_h^e, two intermediate divergence peaks, at B_l^i and B_h^i, and an intense, symmetrical central line, at B_0.[55] When the sample temperature increases above 130 K, the rigid-limit ESR spectrum changes its overall lineshape as a result of the dynamic line broadenings and shifts as well as the motional averaging of the anisotropic part of the inhomogeneous line broadening. Thus, the

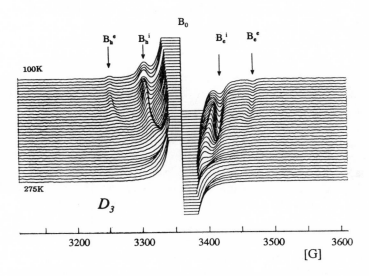

Figure 14. The X-band first-derivative ESR spectra observed for the pure D_3 isomer of dendrimer **5** (G = 1) in benzene at various temperatures.

two extreme steps and the two divergence peaks of the fine-structure spectrum shift toward the central line decreasing the separations between these spectroscopic features, i.e., $\delta B_\| = |B_h^e - B_l^e|$ and $\delta B_\perp = |B_h^i - B_l^i|$. These changes occur in the temperature range 130–195 K and are typical of the lineshape effects in the so-called *slow-tumbling motional region*, where the correlation times are larger than the spin–spin relaxation times, T_2, i.e., $\tau \gg T_2$. At higher temperatures, the dynamic line broadening and line shift are such that the extreme steps are largely masked by the divergence peaks and the last ones are severely overlapped to the central line. It can also be seen from Figure 14 that the extreme steps shift at a faster rate toward the central line than the divergence peaks while the linewidth of the central line gradually decreases. Consequently, in the temperature range from 200 to 250 K, the observed spectra consist of an intense line with two structured shoulders at both sides, the resolution of their shoulders becoming poorer as the sample temperature increases. This temperature range represents the *intermediate-tumbling motional region* where $\tau \approx T_2$. In the vicinity of 250 K, the spectra display a single, symmetric line due to complete motional shifting the spectral features, indicating the beginning of the *fast-tumbling motional region* where $\tau \ll T_2$. At high enough temperatures in this region, the motionally independent, anisotropic contribution to the inhomo-geneous line broadening should vanish with the residual inhomogeneous linewidth corresponding to the motionally independent, isotropic contribution. Thus the spectra in this region, not depicted in Figure 14, consist of a unique symmetric, single line in which its peak-to-peak linewidth decreases with increasing tempera-ture due to dynamic narrowing by the fast-tumbling motion.

Lee et al. evolved a comprehensive analytical–theoretical treatment, based on the solution of the reorientational isotropic diffusion equation, for an ensemble of high-spin systems under motion.[54] These authors developed an analytical expres-sion for the slow-tumbling motional region that relates the orientational-motion correlation time τ (in s), or the corresponding tumbling rate τ^{-1}, with the step separation $\delta B_\|$ of the ESR fine structure of a quartet by Eq. 8,

$$\tau = 1.92756 \cdot 10^{-7} D'/[4D' - \delta B_\|]^2 \qquad (8)$$

where D' is the zero-field splitting parameter of the quartet in Gauss. Up to now we have used such an equation to calculate the tumbling rates of the stereoisomers C_2 and D_3 of dendrimer **5** (G = 1) in benzene and *n*-hexane in the temperature range corresponding to the slow-tumbling motional region. The plots of $\ln(\tau^{-1})$ versus $1/T$ permitted calculation of the activation energies, E_a, corresponding to the tumbling process of both isomers, by means of the Arrhenius equation. The resulting values for the activation energy were 34 ± 2 and 24 ± 1 kJmol^{-1} in benzene and 29 ± 3 and 20 ± 1 kJmol^{-1} in *n*-hexane for the C_2 and D_3 isomers, respectively. The relative magnitudes of these activation energies are very interesting since they provide an indication of the solute/solvent interactions that dominate this dynamic process. In fact, the E_a values in benzene are larger than in *n*-hexane for both isomers, and the values for the D_3 isomer are always larger than those obtained for the C_2 one, regardless the nature of the solvent. Both trends agree with those theoretically expected if the reorientational motions of quartets inside the cavities are mainly limited by the polar interactions since the polarity of benzene is higher than that of n-hexane and the dipolar moment of the C_2 is larger than that of the D_3 (see Table 1). Therefore, in absence of any cavitational effects, the polarity effects control the different tumbling rates that show the two conformers of this high-spin dendrimer.

IV. CONCLUSIONS AND PERSPECTIVES

Dendritic molecular architectures have several advantages from the physical and synthetic points of view with respect to other architectures in order to prepare *polymeric organic magnetic materials*. They are compatible with the most efficient ways to achieve robust intramolecular ferromagnetic interactions and, at the same time, permit us to obtain macromolecules with an extremely large number of coupled open-shell units using only a limited number of synthetic steps. Nevertheless due to the high degree of branch-cell hierarchy, such dendritic architectures have a prominent weakness which is associated with the presence of structural and/or magnetic coupling defects and limits the sizes of the effective S values of the resulting *super high-spin macromolecules*. Practical solutions to this drawback are still awaiting the development of new alternative efficient synthetic procedures that circumvent the consequences of these magnetic defects.

The existence of the studied high-spin dendrimers in several isolable and stable stereoisomeric forms that show striking differences in their overall sizes and shapes as well as in their external surfaces have permitted us to study the influence of certain molecular structural parameters, such as surface area, volume, and fractal dimensionality in the ability of dendritic macromolecules to interact with neighboring solvent molecules. These studies have been performed using the *linear solvation energy relationship* methodology and have demonstrated that the cavity formation process in the solvents is the mechanism that controls the distinct chromatographic retentions of all their stereoisomeric forms; therefore, the distinct sizes and shapes are the unique molecular characteristic responsible for their discrimination. Another study of this kind has revealed that the relative stability and populations of these stereoisomers in solution are controlled first by the differences in their shapes and sizes, and second by their distinct polarities. The last electronic characteristics also rule the tumbling rates in solution of the distinct conformers of this family of dendrimers, since in this case the cavitational effects cannot influence such processes. From all these conclusions, it is expected that the intrinsic fractal character of dendritic macromolecules will play a major role in most of their physicochemical properties, which are worthy of detailed study.

ACKNOWLEDGMENTS

We gratefully acknowledge the Comisión Interministerial de Ciencia y Tecnología (Grant, MAT94-0797), Spain, and the New Energy and Industrial Technology Development (NEDO Grant, *Organic Magnets*), Japan, for support of this research. Acknowledgment is also made to the CIRIT for a partial support of this work (Grant, GRQ93-028). N. V. and D. R. thank the CICyT and Generalitat de Catalunya for the doctoral fellowships.

REFERENCES AND NOTES

1. Tomalia, D. A.; Naylor, A. M.; Goddard III, W. A. *Angew. Chem.* **1990**, *102*, 119; *Angew. Chem. Int. Ed. Engl.* **1990**, *29*, 138.
2. Fréchet, J. M. J. *Science* **1994**, *263*, 1710.
3. Tomalia, D. A.; Durst, H. D. *Top. Curr. Chem.* **1993**, *165*, 193.
4. Gitsov, I.; Wooley, K. L.; Hawker, C. J.; Ivanova, P. T.; Fréchet, J. M. J. *Macromolecules* **1993**, *26*, 5621.
5. Jansen, J. F. G. A.; de Brabander-ven den Berg, E. M. M.; Meijer, E. W. *Science* **1994**, *266*, 1226.
6. Buhleier, E.; Wehner, W.; Vögtle, F. *Synthesis* **1978**, 155.
7. Denkewalter, R. G.; Kolc, J.; Luskasavage, W. J. U.S. Patent 4 289 872, 1983; Aharoni, S. M.; Crosby III, C. R.; Walsh, E. K. *Macromolecules* **1982**, *15*, 1093.
8. Newkome, G. R.; Yao, Z.; Baker, G. R.; Gupta, V. K. *J. Org. Chem.* **1985**, *50*, 2004; Newkome, G. R.; Yao, Z.; Baker, G. R.; Gupta, V. K.; Russo, P. S.; Saunders, M. J. *J. Am. Chem. Soc.* **1986**, *108*, 849.
9. Hawker, C. J.; Fréchet, J. M. J. *J. Am. Chem. Soc.* **1990**, *112*, 7638; *J. Chem. Soc., Chem. Commun.* **1990**, 1010; *Macromolecules* **1990**, *23*, 4726.
10. Miller, T. M.; Neenan, T. X. *Chem. Mater.* **1990**, *2*, 346.
11. For recent overviews on molecular magnetism, see: (a) Gatteschi, D.; Kahn, O.; Miller, J. S.; Palacio, F., Eds. *Molecular Magnetic Materials*; Kluwer Academic Publishers: Dordrecht, 1991; (b) Iwamura, H.; Miller, J. S., Eds.; *Proceedings of International Symposium on Chemistry and Physics of Molecular Based Magnetic Materials*, Tokyo, Japan, October 1992. In *Mol. Cryst. Liq. Cryst.* **1993**, *232*, 25–30; (c) Miller, J. S.; Epstein, A. J. *Angew. Chem. Int. Ed. Engl.* **1994**, *33*, 385; (d) Kahn, O. *Molecular Magnetism*; VCH Publishers: New York, 1993; (e) Miller, J. S., Ed. *Proceedings of International Symposium on Chemistry and Physics of Molecular Based Magnetic Materials*, Salt Lake City, October 1994. In *Mol. Cryst. Liq. Cryst.* **1995**, 16–21.
12. Veciana, J. In *Localized and Itinerant Molecular Magnetism*; Coronado, E., Ed.; Kluwer Academic Publishers: Dordrecht, in press.
13. (a) Iwamura, H.; Koga, N. *Acc. Chem. Res.* **1993**, *26*, 346–351; (b) Borden, W. T.; Iwamura, H.; Berson, J. A. *ibid.* **1994**, *27*, 109; (c) Borden, W. T.; Davidson, E. R. *J. Am. Chem. Soc.* **1977**, *99*, 4587; (d) Ovchinikov, A. A. *Theor. Chim. Acta* **1978**, *47*, 297; (e) Klein, D. J. *Pure Appl. Chem.* **1983**, *55*, 299.
14. Borden, W. T., Ed. *Diradicals*; Wiley: New York, 1982; Karafilogou, P. *J. Chem. Phys.* **1985**, *82*, 3728; Fang, S.; Lee, M.-S.; Horvat, D. A.; Borden, W. T. *J. Am. Chem. Soc.* **1995**, *117*, 6727; Yoshizawa, K.; Hoffmann, R. *J. Am. Chem. Soc.* **1995**, *117*, 6921; and references therein.
15. Fujita, I.; Teki, Y.; Takui, T.; Kinoshita, T.; Itoh, K.; Miko, F.; Sawaki, Y.; Iwamura, H.; Izuoka, A.; Sugawara, T. *J. Am. Chem. Soc.* **1990**, *112*, 4074; and references therein.
16. (a) Rajca, A. In *Advances in Dendritic Macromolecules*; Newkome, G. R., Ed.; JAI Press: Greenwich, CT, 1994, Vol. 1, pp. 133–168; (b) Rajca, A. *Chem. Rev.* **1994**, *94*, 871; and references therein.
17. Nakamura, N.; Inoue, K.; Iwamura, H. *Angew. Chem. Int. Ed. Engl.* **1993**, *32*, 873; and references therein.

18. There have been some claims of ferromagnetic organic polymers. Nevertheless, they are plagued by problems due to dilute fractions of the magnetically active phases, no well-defined chemical structures, possible presence of extrinsic sources of magnetism, and poor reproducibility both in the preparation and in the magnetic behaviors. Consequently, there has not been any report of a polymer that has passed the confirmation test for intrinsically bulk ferromagnetic behavior. See: Miller, J. S. *Adv. Mater.* **1992**, *4*, 298, 435.

19. (a) Rajca, A.; Utamapanya, S. *Mol. Cryst. Liq. Cryst.* **1993**, *232*, 305; (b) Rajca, A.; Utamapanya, S. *J. Am. Chem. Soc.* **1993**, *115*, 10688.

20. (a) Ventosa, N.; Ruiz, D.; Rovira, C.; Veciana, J. *Mol. Cryst. Liq. Cryst.* **1993**, *232*, 333; (b) Veciana, J.; Rovira, C.; Hernández, E.; Ventosa, N. *An. Quim.* **1993**, *89*, 73.

21. Ballester, M. *Acc. Chem. Res.* **1985**, 380; *Adv. Phys. Org. Chem.* **1989**, *25*, 267; and references therein.

22. Veciana, J.; Rovira, C.; Ventosa, N.; Crespo, M. I.; Palacio, F. *J. Am. Chem. Soc.* **1993**, *115*, 57.

23. Only two and twelve diastereoisomeric forms are respectively depicted in Figure 5 for **5** (G = 1) and in Figure 6 for **6** (G = 1). Each one of these forms has its corresponding enantiomer.

24. Ventosa, N. Ph.D. Thesis. Universitat Ramón Llull, Barcelona, 1996.

25. Gust, D.; Mislow, K. *J. Am. Chem. Soc.* **1973**, *95*, 1535; Mislow, K. *Acc. Chem. Res.* **1976**, *9*, 26.

26. Theoretical computations were performed with the software package CERIUS 2, v. 1.6 from Molecular Simulations Inc.

27. Ruiz, D. Ph.D. Thesis. Universitat Autònoma de Barcelona, 1996.

28. The seventh and ninth more retained peaks are in fact an overlapping of two non resolved peaks (numbers 6, 7 and 8, 9 in Figure 7b), as ascertained by changing the composition of the mobile phase.

29. Rajca, A.; Rajca, S.; Desai, S. R. *J. Am. Chem. Soc.* **1995**, *117*, 806.

30. Nishide, H.; Kaneko, T.; Nii, T.; Katoh, K.; Tsuchida, E.; Yamaguchi, K. *J. Am. Chem. Soc.* **1995**, *117*, 548.

31. Langmuir, I. *Colloid. Symp. Monogr.* **1925**, *3*, 3; Pauling L. *Nature* **1948**, *161*, 707; Miertus, S.; Scrocco, E.; Tomasi, J. *Chem. Phys.* **1981**, *55*, 117; Floris, G.; Tomasi, J. *J. Comp. Chem.* **1989**, *10*, 616.

32. Pierotti, R. A. *Chem. Rev.* **1976**, *76*, 717; Huron, M.-J; Claverie, P. *J. Phys. Chem.* **1972**, *76*, 2123.

33. Bibier, A.; André, J. J.; Veciana, J., unpublished results.

34. Since no torsional conformers are expected for each of these rather rigid isomers, the contributions to entropy are only caused by the distinct molecular rotational degrees of freedom of each isomer considered as a rigid unit; which can be evaluated by symmetry considerations. The symmetry contribution to entropy is $-R \ln \sigma$, where σ is the symmetry number; i.e., the number of undistinguishable positions adopted by the isomer (considered rigid) by simple rotations. Symmetry numbers are $\sigma = 2$ for the C_2 isomer and $\sigma = 6$ for the D_3 one.

35. Richards, F. M. *Annu. Rev. Biophys.* **1977**, *6*, 151.

36. Connolly, M. L. *Science* **1983**, *221*, 709.

37. Pascual-Ahuir, J. L.; Silla, E. *J. Comp. Chem.* **1990**, *11*, 1047; Silla, E.; Tuñon, I.; Pascual-Ahuir, J. L. *ibid.* **1991**, *12*, 1077.

38. Lewis, M.; Rees, D. C. *Science* **1985**, *230*, 1163.

39. Pfeizer, P.; Welz, U.; Wippermann, H. *Chem. Phys. Lett.* **1985**, *113*, 535.

40. Mandelbrot, B. B. *The Fractal Geometry of Nature*; Freeman: San Fransisco, 1983; Avnir, D.; Farin, D.; Pfeifer, P. *Nature* **1984**, *308*, 261.

41. (a) Reichardt, C. *Solvent and Solvent Effects in Organic Chemistry*; VCH: New York, 1990; (b) Reichardt, C. *Chem. Rev.* **1994**, *94*, 2319.

42. Kamlet, M. J.; Taft, R. W. *J. Am. Chem. Soc.* **1976**, *98*, 377, 2886; Kamlet, M. J.; Abboud, J.-L. M.; Taft, R. W. *J. Am. Chem. Soc.* **1977**, *99*, 6027; 8325; Taft, R. W.; Abboud, J.-L. M.; Kamlet, M. J.; Abraham, M. H. *J. Solut. Chem.* **1985**, *14*, 153; Abboud, J.-L. M.; Kamlet, M. J.; Taft, R. W. *Prog. Phys. Org. Chem.* **1981**, *13*, 481.

43. Barton, A. F. M. *Handbook of Solubility Parameters and other Cohesion Parameters*; CRC Press: Boca Raton, FL, 1983.

44. Kamlet, M. J.; Taft, R. W. *Acta Chem. Scand. Part B* **1985**, *B39*, 611; Kamlet, M. J.; Dogherty, R. M.; Abraham, M. H.; Taft, R. W. *Quant. Struct. Act. Relat.* **1988**, *7*, 71; and references therein.

45. Chimielowiec, J.; Sawatzky, H. *J. Chromatogr. Sci.* **1979**, *17*, 245; Melander, W. R.; Chen, B.-K.; Horváth, Cs. *J. Chromatogr.* **1979**, *187*, 167; Snyder, L. R. *J. Chromatogr.* **1979**, *179*, 167.

46. Boehm, R. E.; Martire, D. E.; Amstrong, D. W. *Anal. Chem.* **1988**, *60*, 522.

47. Klumpp, G. W. *Reactivity in Organic Chemistry;* John Wiley; New York, 1982, p. 275.

48. Entropy changes were estimated with Eq. 4 assuming that V_s is equal to the total stationary phase volume existing in the column. Therefore, these values reflect more properly the relative differences in entropies of transfer instead of the standard molar entropies that would require to use the volume of the *active* stationary phase.

49. Sadek, P.; Carr, P. W.; Doherty, R. M.; Kamlet, M. J.; Taft, R. W.; Abraham, M. H. *Anal. Chem.* **1985**, *57*, 2971.

50. In order to model the solvent strength parameters of the ODS we used the *n*-hexane as a model solvent due to its structural similarity with the octadecyl chains of this stationary phase.

51. In the opposite case (i.e., if both isomers would have identical formation molar enthalpies), the equilibrium constant and populations would be determined only by the difference in their entropies (see Table 1) and, accordingly, $K^0 = \exp(+\Delta\Delta S^0/R)=1/3$ (see Eq. 7). Therefore, under such circumstances the equilibrium constant of this entropically controlled process would remain unchanged with the temperature and the nature of the solvent.

52. Tomás, X.; Ventosa, N.; Veciana, J., unpublished results.

53. In the present study this term is senseless since **5** ($G = 1$) has no hydrogen atoms to interact with the solvents which have hydrogen-bond donor abilities.

54. Lee, S.; Tang, S.-Z. *Phys. Rev. B* **1985**, *32*, 2761; Lee, S.; Brown, I. M. *ibid.* **1986**, *34*, 1442.

55. Kothe, G.; Ohmes, E.; Brickmann, J.; Zimmermann, H. *Angew. Chem. Int. Ed. Engl.* **1971**, *10*, 938; Reibish, K.; Kothe, G.; Brickmann, J. *J. Chem. Phys. Lett.* **1972**, *17*, 86; Brickmann, J.; Kothe, G. *J. Chem. Phys.* **1973**, *59*, 2807.

DENDRIMERS BASED ON METAL COMPLEXES

Scolastica Serroni, Sebastiano Campagna,
Gianfranco Denti, Alberto Juris,
Margherita Venturi, and Vincenzo Balzani

Advances in Dendritic Macromolecules
Volume 3, pages 61–113
Copyright © 1996 by JAI Press Inc.
All rights of reproduction in any form reserved.
ISBN: 0-7623-0069-8

I. INTRODUCTION

A driving force of modern chemical research is the need for smaller and more efficient components for electronic, optical, and mechanical applications.[1,2] This interest is currently leading to the development of new types of molecular architectures of nanometric dimensions capable of playing a role in the construction of molecular devices.[3-5]

In this context, much attention is currently devoted to the preparation of highly branched tree-like species,[6] variously called cascade molecules,[7] arborols,[8] or dendrimers.[9] The reasons why such compounds are interesting from a fundamental viewpoint and promising for a variety of applications have been reviewed[10] and highlighted[11] by several authors. Many of the dendrimers obtained so far are organic in nature. This review describes a novel family of dendrimers containing metal ions, prepared in our laboratories over the last few years.

Our aim has been the construction of dendrimers that incorporate in their building blocks specific "pieces of information" such as the capability to absorb and emit visible light and to reversibly exchange electrons.[12-16] To pursue this aim, we have designed a synthetic strategy to build up dendrimers based on luminescent and redox-active transition metal complexes. Species containing 4,[17-21] 6,[22-24] 7,[25] 10,[26,27] 13,[28] and 22[29,30] metal-based units have already been obtained. We will see that, in principle, the developed strategy can be used to prepare dendrimers of larger size, containing a number of different metals and/or ligands in specific sites of the supramolecular architecture.[30-32]

II. BUILDING BLOCKS

A. Metal Complexes

In order to build up dendrimers capable of exhibiting redox activity and light-induced functions, appropriate building blocks have to be used. In the last 20 years, extensive investigations carried out on the photochemical and electrochemical properties of transition metal compounds have shown that Ru(II) and Os(II) complexes of aromatic *N*-heterocycles (Figure 1), e.g., $Ru(bpy)_3^{2+}$ and $Os(bpy)_3^{2+}$ (bpy = 2,2′-bipyridine), exhibit a unique combination of chemical stability, redox properties, excited state reactivity, luminescence, and excited state lifetime.[33-37] Furthermore all these properties can be tuned within rather broad ranges by changing ligands or ligand substituents. Several hundreds of these complexes have been synthesized and characterized, and some of them have been used as sensitizers for the interconversion of light and chemical energy as well as for other purposes. More recently, a great number of polynuclear metal complexes containing Ru(II) and Os(II) have also been prepared and their electrochemical and excited state properties have been investigated.[38,39]

It can be shown that for Ru(II) and Os(II) complexes of aromatic *N*-heterocycles the ground state, the low energy excited states, and the redox forms can be described in a sufficiently approximate way by localized molecular orbital configurations.[40–43] With such an assumption, the various electronic transitions are classified as metal-centered (MC), ligand-centered (LC), and charge transfer (CT: either metal-to-ligand, MLCT, or ligand-to-metal, LMCT), and the oxidation and reduction processes can be considered as metal- and ligand-centered, respectively. These complexes exhibit very intense LC bands in the UV region and intense MLCT bands in the visible. Regardless of the excitation wavelength, the originally populated excited states undergo fast radiationless decay to the lowest triplet ^{3}MLCT, which is luminescent both in rigid matrix at 77 K and in fluid solution at room temperature. The properties of $Ru(bpy)_3^{2+}$, the prototype of this class of compounds, are illustrated in Figure 2.

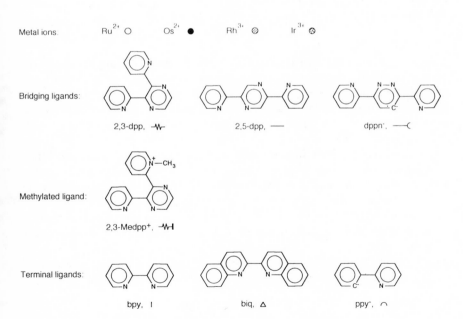

Figure 1. Formulas of the ligands, abbreviations, and symbols used to represent the components of the dendrimers.

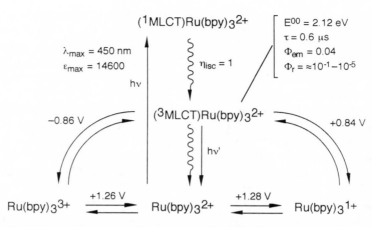

Figure 2. Schematic representation of some relevant ground and excited-state properties of Ru(bpy)$_3^{2+}$. ^{1}MLCT and ^{3}MLCT are the spin-allowed and spin-forbidden metal-to-ligand charge transfer excited states, responsible for the high intensity absorption band with λ_{max} = 450 nm and the luminescence band with λ_{max} = 615 nm, respectively. The other quantities shown are: intersystem crossing efficiency (η_{isc}); energy (E^{00}) and lifetime (τ) of the ^{3}MLCT state; luminescence quantum yield (Φ_{em}); quantum yield for ligand detachment (Φ_r). The reduction potentials of couples involving the ground and the ^{3}MLCT excited states are also indicated.

Most of the luminescent and redox-active metal-based dendrimers studied so far contain N–N chelating aromatic molecules, as ligands, and Ru(II) and Os(II), as metals. Incorporation of other metals in a dendrimer-type structure requires other types of bridging and/or terminal ligands. It is known that stable and photochemically interesting mononuclear complexes of metals like Rh(III), Ir(III), Pd(II), Pt(II), and Pt(IV) can be obtained by using C$^-$–N chelating (cyclometallating) ligands (Figure 1).[44] Therefore, we have used cyclometallating ligands to prepare oligonuclear metal complexes containing some of these metal-based units to prepare dendrimers containing Rh(III) and/or Ir(III).[21,24] It should be pointed out that in cyclometallated complexes, the metal–C$^-$ bonds can exhibit a large degree of covalency and different electronic configurations can lie very close in energy. As a consequence, for cyclometallated complexes the localized molecular orbital approach mentioned above can only be used as a very crude approximation and most often the interpretation of the properties of such complexes is not as easy as for polypyridine complexes.[44]

B. Bridging Ligands

In dendrimers based on metal complexes, the metal-containing units are linked together by bridging ligands. The role played by the bridging ligands is extremely important for the following reasons: (1) their coordinating sites contribute (together with the coordination sites of the "terminal" ligands) to determine the spectroscopic and redox properties of the active metal-based units; (2) their structure and the orientation of their coordinating sites determine the architecture of the dendrimer; and (3) their chemical nature controls the electronic communication between the metal-based units. Therefore the choice of suitable bridging ligands is crucial to obtain dendrimers capable to show luminescence, to exhibit interesting electrochemical properties, and to give rise to photoinduced energy transfer and electron transfer processes.

The bridging ligands, of course, must also satisfy requirements that make the synthetic process efficient and controllable in each step. For example, their coordination sites should not react under the same experimental conditions, otherwise one would mostly obtain compounds of undefined nuclearity (*vide infra*).

In the past few years we have concentrated our efforts on the N–N bischelating bridging ligands 2,3- and 2,5-bis(2-pyridyl)pyrazine (2,3- and 2,5-dpp), respectively (Figure 1). As we will see later, for the synthesis of dendrimers we have mainly used 2,3-dpp since one of its two chelating sites can be protected by methylation.[45] The methylated form of 2,3-dpp, 2,3-Medpp+, is also shown in Figure 1.

More recently we have begun to use the ligand 3,6-bis(2'-pyridyl)pyridazine (dppnH) which, after deprotonation of the pyridazine ring, gives a monoanion (dppn$^-$) capable to coordinate two metal ions on opposite sides by means of a N–N and a C$^-$–N chelating moieties (Figure 1).

III. SYNTHESIS

A. "Complexes as Metals and Complexes as Ligands" Synthetic Strategy

The typical approach used to prepare polynuclear metal complexes is the so-called complexes as metals and complexes as ligands strategy,[12,27] briefly illustrated in the following.

Mononuclear complexes are synthesized by combining metal ion (M) and free ligands (L), as shown in Eq. 1:

$$M + nL \rightarrow M(L)n \tag{1}$$

In the "complexes as metals/complexes as ligands" strategy one uses complexes (building blocks) in the place of the metal (M) and/or of the ligands (L). The place of M can be taken by mono- or oligonuclear complexes that possess easily replaceable ligands, so that they can give rise to species with unsaturated metal coordination sites (complex–metals), and the place of L can be taken by mono- or oligonuclear complexes which contain free chelating sites (complex–ligands).

The application of this strategy to dinuclear complexes is briefly illustrated. If the target is the preparation of a dinuclear homometallic complex, the synthesis is

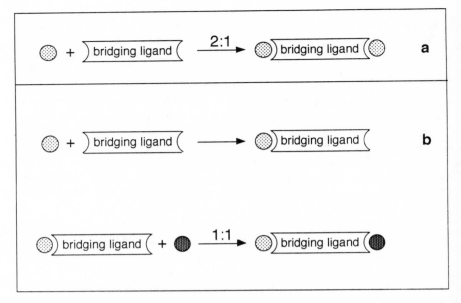

Figure 3. Schematic representation of the synthetic strategy used to prepare dinuclear homometallic (**a**) and dinuclear heterometallic (**b**) complexes.

particularly easy: it is sufficient to react a complex–metal with a bridging ligand in stoichiometric ratio, as illustrated in Figure 3a. To obtain a dinuclear heterometallic complex, the usual approach involves two steps (Figure 3b). First the preparation of a mononuclear complex ligand is performed by reacting a metal precursor with an excess of the ligand. During this step, dinuclear homometallic by-products are unavoidably formed, but the desired mononuclear complex–ligand can be easily isolated by column chromatography (see below). In a following step, the second metal (as complex–metal) is added, so that only the dinuclear heterometallic species can be obtained. In the case of chelate ligands, no scrambling of metals is usually found in the experimental conditions used for the synthesis. To obtain ruthenium–osmium heterometallic complexes of bpy-type bridging ligands, it is convenient to prepare in the first step the mononuclear osmium complex. There are at least two advantages in choosing this procedure: (1) as osmium is less reactive than ruthenium, in the first step the yield of unwanted dinuclear homometallic compounds is minimized; and (2) the major reactivity of the ruthenium complex–metals allows the use of milder conditions to obtain the dinuclear species.

Polynuclear complexes of higher nuclearity can be obtained with this strategy if suitable polynuclear building blocks are available. An example is illustrated in Figure 4 where a mononuclear complex–ligand having three free chelating sites is reacted with three equivalents of a trinuclear complex–metal, thus forming a

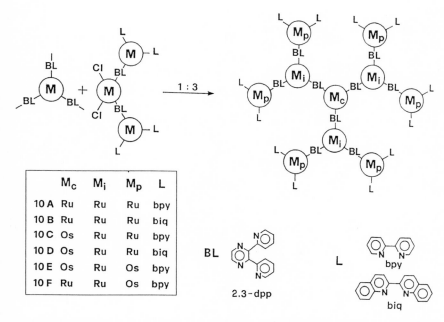

	M_c	M_i	M_p	L
10 A	Ru	Ru	Ru	bpy
10 B	Ru	Ru	Ru	biq
10 C	Os	Ru	Ru	bpy
10 D	Os	Ru	Ru	biq
10 E	Os	Ru	Os	bpy
10 F	Ru	Ru	Os	bpy

Figure 4. Schematic representation of the synthesis of decanuclear compounds.[27]

decanuclear species (convergent synthesis). By a clever choice of the reaction partners it is possible to obtain compounds where different metals and ligands can be located in the desired position of the supramolecular structure. This approach is a typical example of modular chemistry.[46]

Dinuclear cyclometallated complexes are usually prepared starting from the metal chloride and the cyclometallating ligand (Figure 5, top). The same type of synthesis can be used to prepare cyclometallated complexes of high nuclearity. As an example, the recently reported synthesis of a hexanuclear cyclometallated complex is shown in Figure 5, bottom.[21]

B. Protection and Deprotection of the Bridging Ligands

The usual procedure to obtain organic dendrimers is a divergent iterative approach (*vide infra*) which requires the availability of a bifunctional species. In the field of coordination chemistry, this species has to be a complex capable to behave both as a ligand and as a metal. A simple example is the compound [Ru(2,3-dpp)$_2$Cl$_2$], shown in Figure 6. Unfortunately a species like this is unavoidably self-reactive under the preparative conditions because the free chelating sites of one molecule would substitute the labile ligands on another molecule, leading to

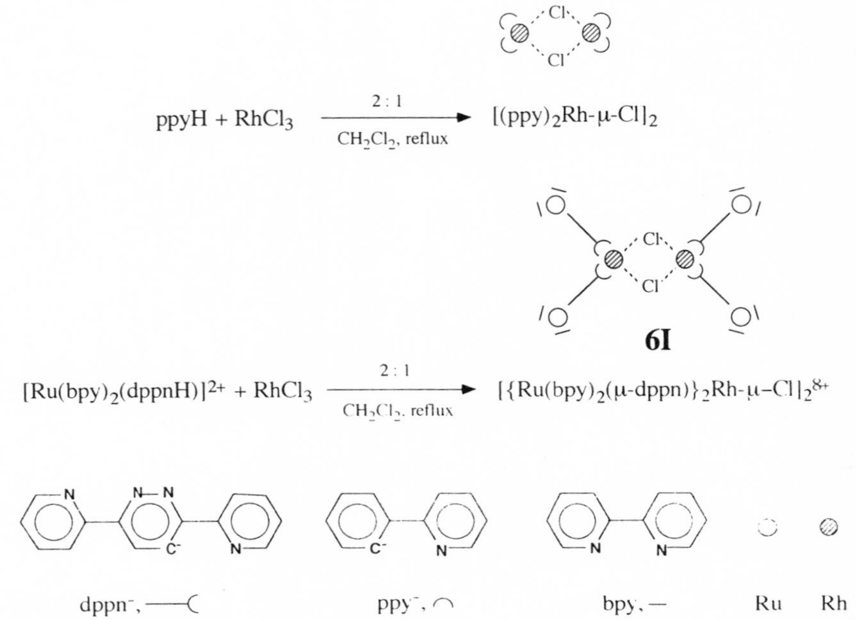

$$ppyH + RhCl_3 \xrightarrow[\text{CH}_2\text{Cl}_2,\ \text{reflux}]{2:1} [(ppy)_2Rh\text{-}\mu\text{-Cl}]_2$$

6I

$$[Ru(bpy)_2(dppnH)]^{2+} + RhCl_3 \xrightarrow[\text{CH}_2\text{Cl}_2,\ \text{reflux}]{2:1} [\{Ru(bpy)_2(\mu\text{-dppn})\}_2Rh\text{-}\mu\text{-Cl}]_2^{8+}$$

dppn⁻, ——⟨   ppy⁻, ⌒   bpy, —   Ru   Rh

Figure 5. (a) Metal chlorides can react with simple cyclometalating ligands to give dinuclear cyclometallated complexes. (b) Schematic representation of the synthesis of a hexanuclear complex containing cyclometallated units.[21]

compounds of uncontrolled nuclearity. The only safe way to carry on a divergent synthesis must be based on species where one of the two functions is temporarily blocked. This has been achieved by methylation of 2,3-dpp at one pyridyl nitrogen[45] by using trimethyloxonium tetrafluoroborate or methyl triflate (methyl trifluoromethanesulfonate) as alkylating agents to obtain the "protected" ligand 2,3-Medpp⁺. This allows the easy, high-yield (90%) preparation of the complex–metal $[Ru(2,3\text{-Medpp})_2Cl_2]^{2+}$, where the complex–ligand ability is inhibited (Figure 6). The protection is stable under the conditions employed for the successive reactions of $[Ru(2,3\text{-Medpp})_2Cl_2]^{2+}$ as a complex–metal. Furthermore, it has been possible to set up a demethylation procedure fully compatible with the stability of coordinate bonds in order to restore the presence of free chelating sites—that is, the complex–ligand ability.

C. Divergent Synthetic Approach

Recently the divergent iterative approach, well known in organic synthesis,[7–9] has been successfully extended to prepare polynuclear dendrimer-type complexes containing up to 22 metal centers using 2,3-dpp as bridging ligand.[29,30] The two

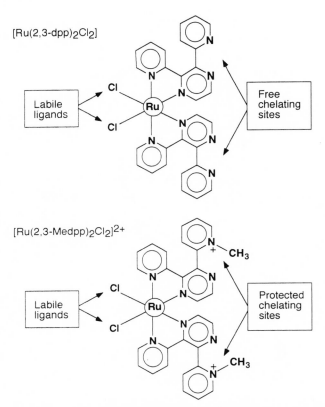

Figure 6. Schematic representation of the bifunctional [Ru(2,3-dpp)$_2$Cl$_2$] and monofunctional [Ru(2,3-Medpp)$_2$Cl$_2$]$^{2+}$ complexes.

key compounds for the synthesis of our dendrimers are the mononuclear [Ru(2,3-dpp)$_3$]$^{2+}$ and [Ru(2,3-Medpp)$_2$Cl$_2$]$^{2+}$ species. The former compound is a complex–ligand since it contains three vacant chelating sites. This species plays the role of a "core" in the synthesis of our dendrimers. As mentioned above (Figure 6), [Ru(2,3-Medpp)$_2$Cl$_2$]$^{2+}$ contains two labile Cl$^-$ ions, and therefore it can play the role of a complex–metal. Once it has been used as a complex–metal, its two methylated ligands can be deprotected, and the new compound can thus be used as a complex–ligand.

The divergent synthetic approach is schematically summarized in Figure 7. Reaction of the mononuclear complex–ligand [Ru(2,3-dpp)$_3$]$^{2+}$ with the complex–metal [Ru(2,3-Medpp)$_2$Cl$_2$]$^{2+}$ in a 1:3 molecular ratio leads to a tetranuclear complex (first dendrimer generation) which in its periphery contains six protected chelating sites. Demethylation of this species restores its ligand ability and yields

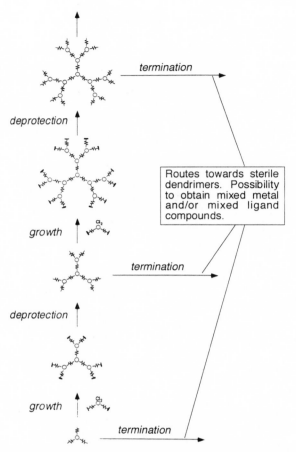

Figure 7. Divergent synthetic strategy to obtain polynuclear metal complexes of dendrimer shape.[29,30] Each deprotected compound of the divergent synthetic approach can be used as a core in convergent synthetic processes. Some of these routes starting from a tetranuclear core are illustrated in Figure 8.

a tetranuclear complex which can play the role of complex–ligand. Further reaction with $[Ru(2,3\text{-Medpp})_2Cl_2]^{2+}$ in 1:6 molecular ratio leads to a protected decanuclear compound (second dendrimer generation). Deprotection of this complex leads to a decanuclear complex–ligand which contains 12 vacant chelating sites. The divergent iterative synthesis has been carried out up to this stage. In principle, iteration of this procedure can lead to further generations, but difficulties are to be expected due to the increasing number of reaction sites.[10] This iterative synthetic strategy is characterized by a *full, step-by-step control of the growing process.* Therefore, different building blocks containing different metals and/or ligands can be introduced at each step. Moreover, each deprotected compound of the divergent synthetic approach can be used as a ligand core in convergent synthetic processes with complex–metals carrying terminal ligands to yield *sterile* dendrimers of higher generation.[27,30]

Figure 8 shows schematically some of the compounds that have been[27] or can be obtained using the tetranuclear $[\{(2,3\text{-dpp})_2Ru(\mu\text{-}2,3\text{-dpp})\}_3Ru]^{8+}$ complex–ligand core.

A complete list of the polynuclear compounds of this family that we have prepared with the synthetic routes used is given in Table 1.[47] The schematic representation of all the complexes is given in Scheme 1.

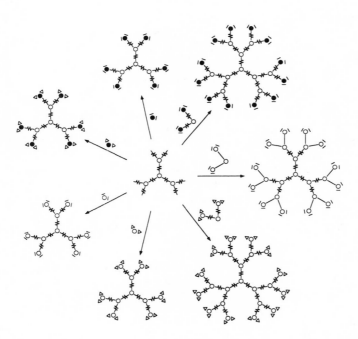

Figure 8. As shown in Figure 7, each deprotected compound encountered in the divergent synthetic approach can be used as a core for convergent synthetic processes. This figure shows schematically some of the compounds that have been[27] or can be obtained using the tetranuclear complex ligand core. The meaning of the symbols used is as follows: O, Ru^{2+}; ●, Os^{2+}; —W—, 2,3-dpp bridging ligand; —, 2,5-dpp bridging ligand; I, bpy terminal ligand; Δ, biq terminal ligand. For the sake of simplicity, in the formulas of the complex metals that react with the tetranuclear core the two labile chloride ligands have been omitted.

Table 1. Compounds Prepared and Synthetic Routes Used[a]

Dinuclear Compounds

method (i)

$$M_a(L_a)_2Cl_2 + [M_b(L_b)_2(BL)]^{2+} \xrightarrow{1:1} [(L_a)_2M_a(\mu\text{-}BL)M_b(L_b)_2]^{4+}$$

2A [18]	$M_a = M_b = Ru$	BL = 2,3-dpp	$L_a = L_b = bpy$
2B [18]	$M_a = M_b = Ru$	BL = 2,3-dpp	$L_a = L_b = biq$
2C [18]	$M_a = M_b = Ru$	BL = 2,3-dpp	$L_a = bpy, L_b = biq$
2D [18]	$M_a = M_b = Ru$	BL = 2,5-dpp	$L_a = L_b = bpy$
2E [18]	$M_a = M_b = Ru$	BL = 2,5-dpp	$L_a = L_b = biq$
2F [18]	$M_a = M_b = Ru$	BL = 2,5-dpp	$L_a = bpy, L_b = biq$
2G [57]	$M_a = Ru, M_b = Os$	BL = 2,3-dpp	$L_a = L_b = bpy$
2H	$M_a = Ru, M_b = Os$	BL = 2,5-dpp	$L_a = L_b = bpy$
2I	$M_a = Ru, M_b = Os$	BL = 2,5-dpp	$L_a = bpy, L_b = biq$

method (ii)

$$[M(L)_2Cl_2]^{2+} + BL \xrightarrow{2:1} [(L)_2M(\mu\text{-}BL)M(L)_2]^{8+}$$

2J [47c]	M = Os	BL = 2,3-dpp	L = bpy
2K [47c]	M = Os	BL = 2,5-dpp	L = bpy
2L [47d]	M = Ru	BL = 2,3-dpp	L = 2,3-Medpp$^+$

Trinuclear Compounds

method (i)

$$M_a(L_a)_2Cl_2 + [M_b(L_b)(BL)_2]^{2+} \xrightarrow{2:1} [(L_a)_2M_a(\mu\text{-}BL)M_b(L_b)(\mu\text{-}BL)M_a(L_a)_2]^{6+}$$

3A [47b]	$M_a = M_b = Ru$	BL = 2,3-dpp	$L_a = L_b = bpy$
3B [47b]	$M_a = M_b = Ru$	BL = 2,3-dpp	$L_a = biq, L_b = bpy$
3C [47b]	$M_a = M_b = Ru$	BL = 2,5-dpp	$L_a = L_b = bpy$
3D [47b]	$M_a = M_b = Ru$	BL = 2,5-dpp	$L_a = biq, L_b = bpy$
3E	$M_a = Os, M_b = Ru$	BL = 2,3-dpp	$L_a = L_b = bpy$
3F	$M_a = Os, M_b = Ru$	BL = 2,5-dpp	$L_a = L_b = bpy$

method (ii)

$$RuCl_3 + [M(L)_2(BL)]^{2+} \xrightarrow{1:2} [Ru\{(\mu\text{-}BL)M(L)_2\}_2Cl_2]^{4+}$$

3G [19,22]	M = Ru	BL = 2,3-dpp	L = bpy
3H [27]	M = Ru	BL = 2,3-dpp	L = biq
3I [22]	M = Ru	BL = 2,5-dpp	L = bpy
3J [27]	M = Os	BL = 2,3-dpp	L = bpy

Tetranuclear Compounds

method (i)

$$M_a(L)_2Cl_2 + [M_b(BL)_3]^{2+} \xrightarrow{3:1} [M_b\{(\mu\text{-}BL)M_a(L)_2\}_3]^{8+}$$

4A [18]	$M_a = M_b = Ru$	BL = 2,3-dpp	L = bpy
4B [18]	$M_a = M_b = Ru$	BL = 2,3-dpp	L = biq
4C [17]	$M_a = Ru, M_b = Os$	BL = 2,3-dpp	L = bpy

method (ii)

$$M_aCl_3 + [M_b(L)_2(BL)]^{2+} \xrightarrow{1:3} [M_a\{(\mu\text{-}BL)M_b(L)_2\}_3]^{8+}$$

4D [57]	$M_a = M_b = Ru$	BL = 2,5-dpp	L = bpy
4E [57]	$M_a = M_b = Ru$	BL = 2,5-dpp	L = biq
4F [57]	$M_a = Ru, M_b = Os$	BL = 2,5-dpp	L = bpy

(continued)

Table 1. (Continued)

method (iii)

$[M_a\{(\mu\text{-}BL_a)M_a(L_a)_2\}_2Cl_2]^{4+} + [M_b(L_b)_2(BL_b)]^{2+} \xrightarrow{1:1} [\{(L_a)_2M_a(\mu\text{-}BL_a)\}_2M_a(\mu\text{-}BL_b)M_b(L_b)_2]^{8+}$

4G [19]	$M_a = M_b = Ru$	$BL_a = 2,3\text{-dpp},$ $BL_b = 2,5\text{-dpp}$	$L_a = L_b = bpy$
4H [20]	$M_a = M_b = Ru$	$BL_a = BL_b = 2,3\text{-dpp}$	$L_a = bpy, L_b = biq$
4I [20]	$M_a = Ru, M_b = Os$	$BL_a = BL_b = 2,3\text{-dpp}$	$L_a = L_b = bpy$

method (iv)

$[M_a(L)_2Cl_2]^{2+} + [M_b(BL)_3]^{2+} \xrightarrow{3:1} [M_b\{(\mu\text{-}BL)M_a(L)_2\}_3]^{14+}$

4J [30]	$M_a = M_b = Ru$	$BL = 2,3\text{-dpp}$	$L = 2,3\text{-Medpp}^+$
4K [16]	$M_a = Ru, M_b = Os$	$BL = 2,3\text{-dpp}$	$L = 2,3\text{-Medpp}^+$

method (v)

$[M\{(\mu\text{-}BL)M(L)_2\}_3]^{14+} \xrightarrow{deprotection} [M\{(\mu\text{-}BL)M(BL)_2\}_3]^{8+}$

4L [30]	$M = Ru$	$BL = 2,3\text{-dpp}$	$L = 2,3\text{-Medpp}^+$

method (vi)

$[M_a(L)_2(\mu\text{-}Cl)]_2 + [M_b(BL)_3]^{2+} \xrightarrow{1.5:1} [M_b\{(\mu\text{-}BL)M_a(L)_2\}_3]^{5+}$

4M [21]	$M_a = Rh, M_b = Ru$	$BL = 2,3\text{-dpp}$	$L = ppy^-$
4N [21]	$M_a = Ir, M_b = Ru$	$BL = 2,3\text{-dpp}$	$L = ppy^-$
4O [21]	$M_a = Rh, M_b = Os$	$BL = 2,3\text{-dpp}$	$L = ppy^-$
4P [21]	$M_a = Ir, M_b = Os$	$BL = 2,3\text{-dpp}$	$L = ppy^-$

Hexanuclear Compounds

method (i)

$[M_a\{(\mu\text{-}BL_a)M_b(L)_2\}_2Cl_2]^{4+} + BL_b \xrightarrow{2:1} [\{(L)_2M_b(\mu\text{-}BL_a)\}_2M_a(\mu\text{-}BL_b)M_a\{(\mu\text{-}BL_a)M_b(L)_2\}_2]^{12+}$

6A [22]	$M_a = M_b = Ru$	$BL_a = BL_b = 2,3\text{-dpp}$	$L = bpy$
6B [22]	$M_a = M_b = Ru$	$BL_a = BL_b = 2,5\text{-dpp}$	$L = bpy$
6C [22]	$M_a = M_b = Ru$	$BL_a = 2,3\text{-dpp},$ $BL_b = 2,5\text{-dpp}$	$L = bpy$
6D [22]	$M_a = M_b = Ru$	$BL_a = 2,5\text{-dpp},$ $BL_b = 2,3\text{-dpp}$	$L = bpy$
6E [23]	$M_a = M_b = Ru$	$BL_a = BL_b = 2,3\text{-dpp}$	$L = biq$
6F [23]	$M_a = Ru, M_b = Os$	$BL_a = BL_b = 2,3\text{-dpp}$	$L = bpy$

method (ii)

$[M_a\{(\mu\text{-}BL_a)M_b(L_a)_2\}_2Cl_2]^{4+} + [M_a(BL_a)\{(\mu\text{-}BL_b)M_a(L_b)_2\}_2]^{6+} \xrightarrow{1:1}$
$[\{(L_a)_2M_b(\mu\text{-}BL_a)\}_2M_a(\mu\text{-}BL_a)M_a\{(\mu\text{-}BL_b)M_a(L_b)_2\}_2]^{12+}$

6G [23]	$M_a = M_b = Ru$	$BL_a = 2,3\text{-dpp},$ $BL_b = 2,5\text{-dpp}$	$L_a = biq, L_b = bpy$
6H [23]	$M_a = Ru, M_b = Os$	$BL_a = 2,3\text{-dpp},$ $BL_b = 2,5\text{-dpp}$	$L_a = L_b = bpy$

method (iii)

$M_aCl_3 + [M_b(L_a)_2(L_b)]^{2+} \xrightarrow{1:2} [\{M_b(L_a)_2(\mu\text{-}BL)\}_2M_a(\mu\text{-}Cl)]_2^{8+}$

6I [24]	$M_a = Rh, M_b = Ru$	$BL = dppn^-$	$L_a = bpy, L_b = dppnH$

Heptanuclear Compounds

$M(L)_2Cl_2 + [M\{(\mu\text{-}BL)M(L)(BL)\}_3]^{8+} \xrightarrow{3:1} [M\{(\mu\text{-}BL)M(L)(\mu\text{-}BL)M(L)_2\}_3]^{14+}$

7A [25]	$M = Ru$	$BL = 2,3\text{-dpp}$	$L = bpy$

(continued)

Table 1. (Continued)

Decanuclear Compounds

method (i)

$[M_a\{(\mu\text{-BL})M_b(L)_2\}_2Cl_2]^{4+} + [M_c(BL)_3]^{2+} \overset{3:1}{\longrightarrow} [M_c\{(\mu\text{-BL})M_a[(\mu\text{-BL})M_b(L)_2]_2\}_3]^{20+}$

10A [26,27]	$M_a = M_b = M_c = Ru$	BL = 2,3-dpp	L = bpy
10B [27]	$M_a = M_b = M_c = Ru$	BL = 2,3-dpp	L = biq
10C [27]	$M_a = M_b = Ru, M_c = Os$	BL = 2,3-dpp	L = bpy
10D [27]	$M_a = M_b = Ru, M_c = Os$	BL = 2,3-dpp	L = biq
10E [27]	$M_a = Ru, M_b = M_c = Os$	BL = 2,3-dpp	L = bpy
10F [27]	$M_a = M_c = Ru, M_b = Os$	BL = 2,3-dpp	L = bpy

method (ii)

$[M(L)_2Cl_2]^{2+} + [M\{(\mu\text{-BL})M(BL)_2\}_3]^{8+} \overset{3:1}{\longrightarrow} [M\{(\mu\text{-BL})M[(\mu\text{-BL})M(L)_2]_2\}_3]^{32+}$

10G [30]	M = Ru	BL = 2,3-dpp	L = 2,3-Medpp⁺

method (iii)

$[M\{(\mu\text{-BL})M[(\mu\text{-BL})M(L)_2]_2\}_3]^{32+} \overset{deprotection}{\longrightarrow} [M\{(\mu\text{-BL})M[(\mu\text{-BL})M(BL)_2]_2\}_3]^{20+}$

10H [30]	M = Ru	BL = 2,3-dpp	L = 2,3-Medpp⁺

Tridecanuclear Compounds

$[M\{(\mu\text{-BL})M(L)_2\}_2Cl_2]^{4+} + [M\{(\mu\text{-BL})M(L)(BL)\}_3]^{8+} \overset{3:1}{\longrightarrow}$
$[M\{(\mu\text{-BL})M(L)(\mu\text{-BL})M[(\mu\text{-BL})M(L)_2]_2\}_3]^{26+}$

13A [28]	M = Ru	BL = 2,3-dpp	L = bpy

Docosanuclear Compounds

$[M\{(\mu\text{-BL})M(L)_2\}_2Cl_2]^{4+} + [M\{(\mu\text{-BL})M(BL)_2\}_3]^{8+} \overset{6:1}{\longrightarrow}$
$[M\{(\mu\text{-BL})M[(\mu\text{-BL})M\{(\mu\text{-BL})M(L)_2\}_2]_2\}_3]^{44+}$

22A [29]	M = Ru	BL = 2,3-dpp	L = bpy

Note: ᵃOriginal references quoted in brackets.

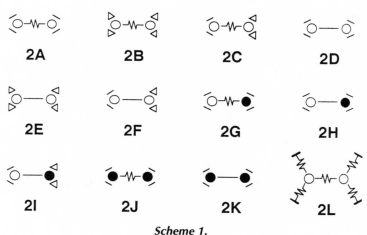

Scheme 1.

Scheme 1. (Continued)

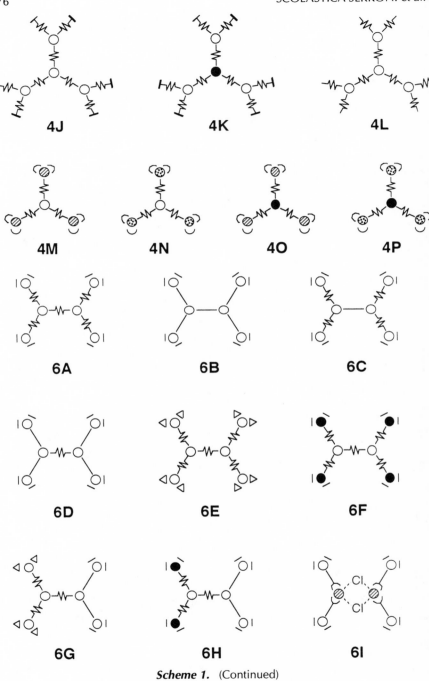

Scheme 1. (Continued)

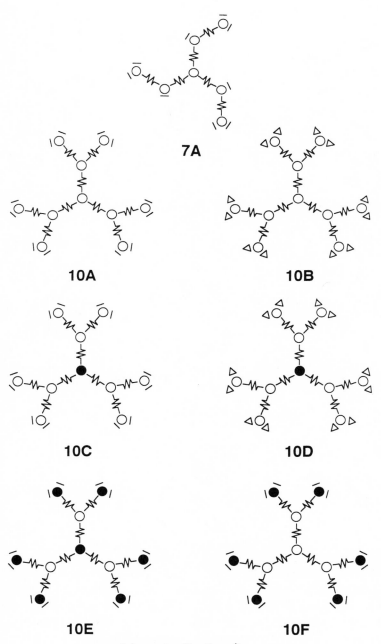

7A

10A   **10B**

10C   **10D**

10E   **10F**

Scheme 1. (Continued)

77

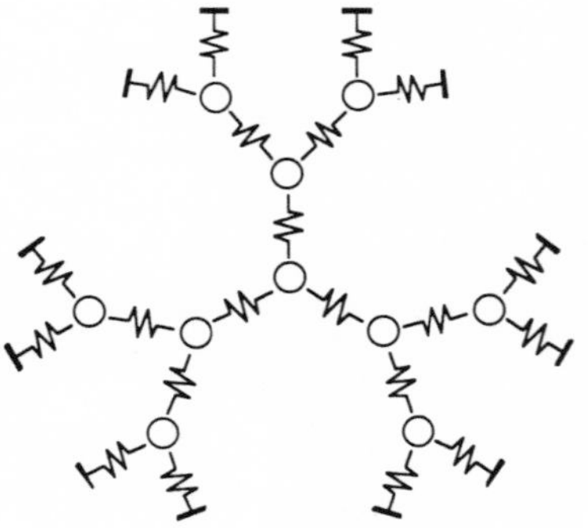

10G

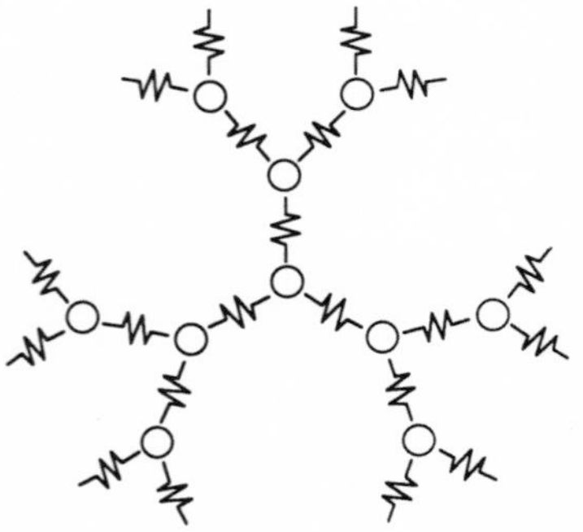

10H

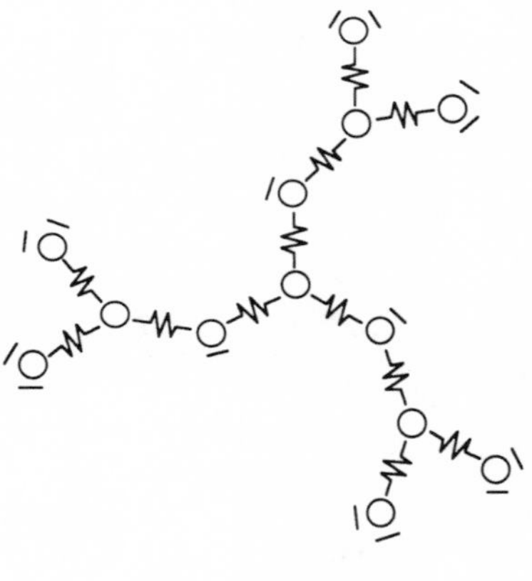

13A

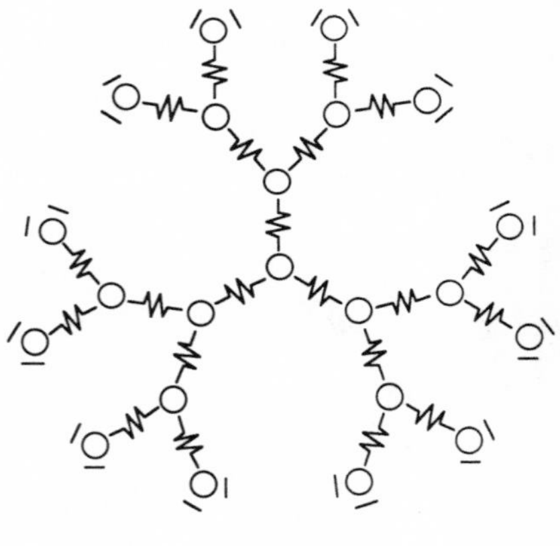

22A

Scheme 1. (Continued)

D. Purification and Characterization

In the synthesis of precursor complex–metal or complex–ligand species, by-products of higher nuclearity are also formed due to the presence in the target molecules of reactive sites (labile ligands or free chelating sites). In the reaction between complex–ligands and complex–metals, the by-products (precursors and partially unreacted species) are lower nuclearity species with respect to the desired products. In both cases, the purification problem consists in the separation of species having different charge and/or size. The obvious answer is chromatography. The choice of the support depends on the desired product. If one needs the species of lower charge and/or smaller size, alumina and silica gel can be used. Ion exchange resins can also be profitably employed for complexes with charge up to 8+. If the target product is the higher charged and/or bigger species, size exclusion resins can be used.

Column chromatographic techniques have been utilized for the most part. The development of analogous HPLC methods[48] will certainly assist in solving the purification problems.

Characterization of large molecules like dendrimers is a difficult task.[10] For neutral compounds, techniques based on colligative properties can be used to determine the molecular mass. Our compounds, however, are highly charged species, and the use of the above techniques is not advisable because of the high number of counterions. Light scattering can hardly be used because of the strong absorption in all the UV and visible spectral region. Mass spectrometry has not yet been extensively used for these compounds, although some results obtained by fast atom bombardment[49] and electrospray ionization[50,51] on smaller polynuclear complexes are promising.

In spite of the above difficulties, a reliable characterization of our dendrimers was achieved by using a variety of techniques. (1) Each compound (including precursors) was purified until TLC showed the presence of only one spot and checked to be stable under the experimental conditions. In particular no ligand scrambling was observed. (2) Each one of three types of steps (growth, deprotection, termination; Figure 7) was accurately controlled as follows:

1. *Growing steps.* (a) The reaction between complex–ligand and complex–metal was carried out under stoichiometric conditions. TLC analysis showed that in each case at least 90% of $[Ru(2,3-Medpp)_2Cl_2]^{2+}$ was reacted. (b) For the product containing protected ligand obtained in each growing step, the ratio between aromatic and aliphatic protons in the ^1H NMR spectrum (where the strong signals of the methyl protons lie in a clean spectral window around 4 ppm) was consistent with the expected formulations. (c) IR analysis on the products containing protected ligands showed the absence of the 990 cm^{-1} band of unbridged 2,3-dpp.

2. *Deprotection steps.* These steps were carried on with a large excess of demethylating agent. The purified products did not show any ^1H NMR signal due to the presence of methyl groups, so we could exclude the presence of residual methylated sites (< 1%).

3. *Termination steps.* As far as these steps are concerned, we have tested the reaction of the complex–ligands species with the complex–metal $[\{Ru(bpy)_2(\mu-2,3-dpp)\}_2RuCl_2]^{4+}$ (which was fully characterized by several techniques including FAB MS).[27,49b] The reaction has been carried out under

stoichiometric conditions until complete ($\geq$ 90%) disappearance of $[\{Ru(bpy)_2(\mu\text{-}2,3\text{-dpp})\}_2RuCl_2]^{4+}$ (TLC analysis). Lack of reactivity toward methylation of the "sterile" products under conditions in which $[Ru(bpy)_2(2,3\text{-dpp})]^{2+}$ is fully methylated demonstrated the absence of free chelating sites (easily checked by ^1H NMR). This shows that the termination step had gone to completion (3% uncertainty).

As we will see below, the luminescence and electrochemical properties are fully consistent with the reported formulations.

The presence of geometrical and optical isomers in complexes containing polypyridine ligands (*vide infra*) makes the structural investigation of these species by NMR technique difficult because the numerous ^1H and ^{13}C signals of the different ligands are multiplied by the number of isomers. To some extent structural information can be obtained from the NMR spectrum of the transition metals, such as ^{99}Ru.[52,53] However, because the ruthenium NMR signals are very broad, the obtainable information is limited in most of the cases to the knowledge of the number of metal ions in the molecule.

More information from NMR analysis can be obtained when the polypyridine ligands carry aliphatic substituents because the NMR signals of such groups are in a clear region of the spectrum and can be easily distinguished.[30]

IV. PROPERTIES

A. Size, Shape, Structure, and General Properties

All the compounds described in this chapter are well-defined supramolecular species, soluble in common solvents (e.g., CH_2Cl_2, CH_3CN), and stable both in the dark and under light excitation.[54] They carry an overall positive charge that, except for the case of cyclometallated and "protected" complexes, is twice the number of the metal atoms.

Schematic representations, like those shown in the Scheme 1, are very useful to indicate the chemical composition of the various species and to discuss the interaction between the various building blocks. Furthermore, as one can understand from the representations shown in Figures 9 and 10, the species with high nuclearity exhibit a three-dimensional branching structure of the type of those shown by otherwise completely different dendrimers based on organic components.[10] Therefore, endo- and exo-receptor properties can be expected, which are currently under investigation. Furthermore, aggregation of decanuclear complexes has also been demonstrated by dynamic light-scattering and conductivity experiments.[55] Systematic studies on aggregation properties have, however, not been performed yet.

The compounds with high nuclearity are indeed very extended and complex structures. Not counting the 44 PF_6^- counterions, the species with 22 metals[30] is made of 1090 atoms, has a molecular weight of 10890 daltons, and an estimated size of 5 nm. Besides the 22 Ru(II) metal atoms, it contains 24 terminal bpy ligands and 21 2,3-dpp bridging ligands.

In principle, these compounds can exist as different isomers because the two coordinating nitrogen atoms of each chelating site of the bridging ligands are not equivalent. A 2D-COSY 400 MHz ^{1}H NMR spectrum of [Ru(2,3-dpp)$_3$]$^{2+}$, which is the "core" in most of our dendrimers, has shown that the purified material is a mixture of the *mer* and *fac* isomers in which the *mer* isomer predominates (92%).[53] The polymetallic complexes can also be a mixture of several diastereoisomeric species since each metal center is also a stereogenic center. For these reasons structural investigations are difficult. Differences arising from the presence of isomeric species, however, are not expected to be large in the electrochemical and spectroscopic properties described below.[56]

We would like to stress that our dendrimers differ from most of those prepared so far for two fundamental reasons: (1) each metal-containing unit exhibits valuable intrinsic properties such as absorption of visible (solar) light, luminescence, and oxidation and reduction levels at accessible potentials; (2) by a suitable choice of the building blocks, different metals and/or ligands can be placed in specific sites of the supramolecular array, as we have already shown for the tetra-,[17-21] hexa-,[22-24] and decanuclear[26,27,30] species. In other words, our dendrimers are species which can incorporate many "pieces of information" and therefore can be used to perform valuable functions such as light harvesting, directional energy transfer, and exchange of a controlled number of electrons at a certain potential.[13-19,22-30,57-61]

Some important properties of the [M(L)$_n$(BL)$_{3-n}$]$^{2+}$ units (M = Ru^{2+} or Os^{2+}; L = bpy or biq; BL = 2,3- or 2,5-dpp) can be summarized as follows.[12,14,57,61] They show (1) intense LC absorption bands in the UV region and moderately intense (ε_{max} ~1 × 10^4 M^{-1}cm^{-1}) MLCT bands in the visible region; (2) a relatively long-lived luminescence (10^{-7}–10^{-8} s at room temperature) in the red spectral region; (3) reversible one-electron oxidation of the metal ion in the potential window +0.8/+1.7 V (vs. SCE); and (4) reversible one-electron reduction of each ligand in the potential window −0.6/−1.1 V (vs. SCE).

Other important differences relevant to our discussion are: (1) Os(II) complexes are oxidized at potentials considerably less positive than Ru(II) complexes; (2) the MLCT absorption and luminescence bands lie at lower energies for the Os(II) complexes than for the Ru(II) ones; (3) the energy of the LUMO of the (monocoordinated) ligands decreases in the series bpy > 2,3-dpp > 2,5-dpp > biq; as a consequence, the lowest (luminescent) ^{3}MLCT level involves the lowest ligand of the above series which is present in the complex; and (4) the electron donor power decreases in the ligand series bpy > biq ≥ 2,3-dpp ~ 2,5-dpp.

The Rh(III) and Ir(III) cyclometallated units present in some of our compounds show: (1) intense LC absorption bands in the UV region and weak bands, mainly of MLCT character, in the visible region of the spectrum; (2) irreversible, presumably metal-based oxidation processes; and (3) reversible one-electron reduction of

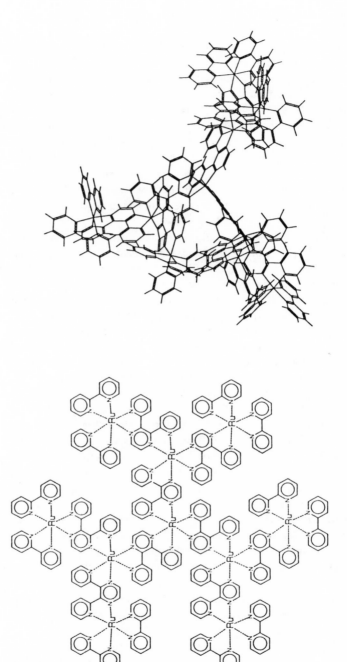

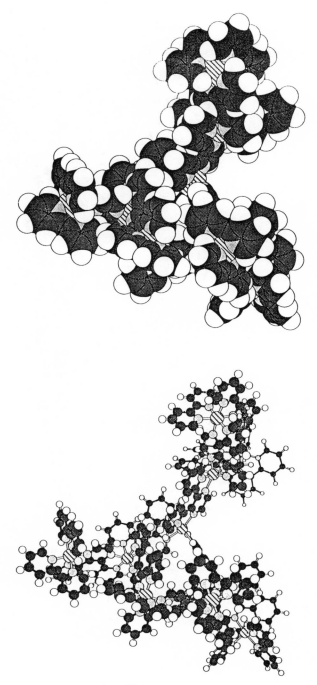

Figure 9. Different schematic representation of a species with 10 metal atoms.

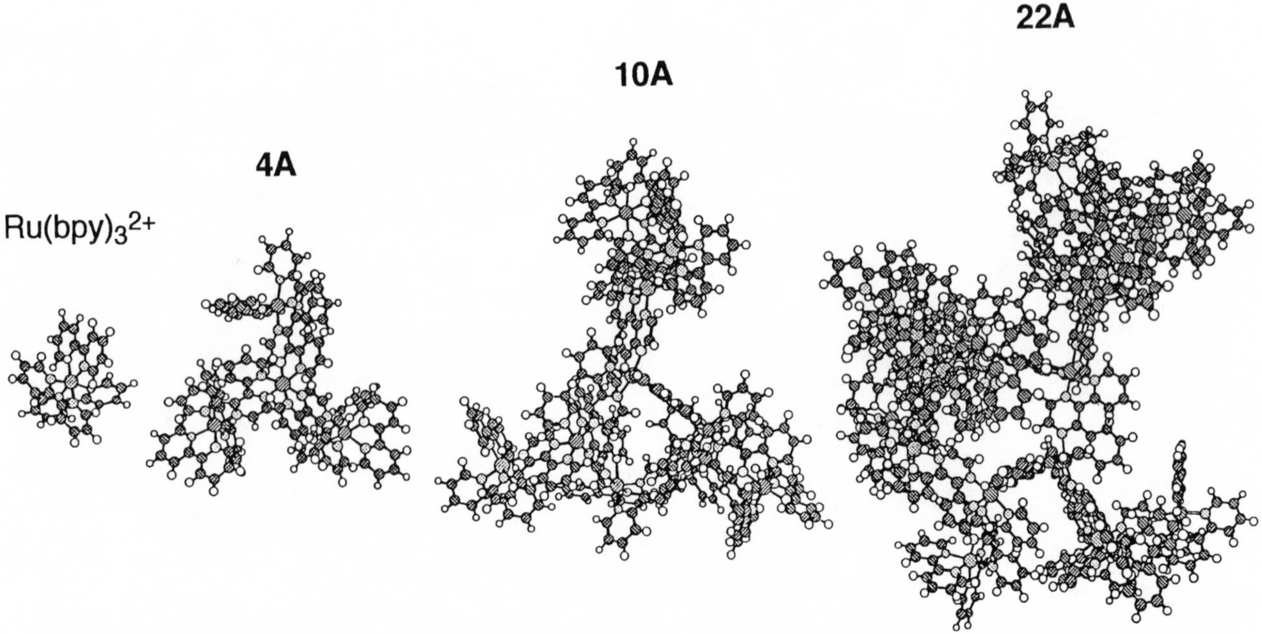

Ru(bpy)$_3^{2+}$

4A

10A

22A

Figure 10. Computer models of [Ru(bpy)$_3$]$^{2+}$, and of three polynuclear complexes containing 4, 10, and 22 metal atoms.

the N–N type ligands which accompany the cyclometallating ligand. More details will be given below.

B. Electrochemical Behavior

Component Units

As mentioned above, the mononuclear compounds of the $[M(L)_n(BL)_{3-n}]^{2+}$ series ($M = Ru^{2+}$ or Os^{2+}; $L = $ bpy or biq; $BL = $ 2,3- or 2,5-dpp) exhibit reversible redox processes. The reduction processes are essentially ligand-localized. Each L ligand is reduced twice and each BL ligand is reduced four times in the potential window $-0.5/-3.1$ V.[62] The reduction potential of each ligand depends on its electronic properties and, to a smaller extent, on the nature of the metal and the other ligands coordinated to the metal. When 2,3- and 2,5-dpp play the role of bis-chelated bridging ligands, their reduction potential becomes more positive by about 0.4 V. Methylation of 2,3-dpp mainly concerns one of the pyridine rings. Therefore a new easily reducible center (the methylated pyridine ring) is created.[47d] In the dendritic species each metal-based unit brings its own redox properties. It should be pointed out, however, that these properties are affected by intercomponent interaction (*vide infra*). On reduction of a polynuclear complex which contains several ligands (e.g., 21 ligands are present in the decanuclear complexes), an extremely complicated pattern with many overlapping peaks is observed.[30] As far as the cyclometallated units are concerned, the presence of anionic ligands displaces the reduction potential to more negative values. For N–N and N–C⁻ mixed-ligand units, reduction of the N–N type ligands occurs first and it is usually reversible, contrary to what happens for the N–C⁻ type moieties.[44]

On oxidation of mononuclear complexes in the potential window $< +1.6$ V, only one metal-based process is observed. The oxidation potential depends strongly on the nature of the metal ion (Os^{2+} is oxidized at less positive potentials compared to Ru^{2+}) and, less dramatically, on the nature of the coordinated ligands. Because of the previously discussed electronic properties of the isolated components and of the stabilization of the LUMO of 2,3-dpp and 2,5-dpp on coordination to a second metal center, it can be expected that the oxidation potential of the metal-containing building blocks increases in the order:[12,14,57]

$$Os(bpy)_2(\mu\text{-}2,5\text{-dpp})^{2+} \leq Os(bpy)_2(\mu\text{-}2,3\text{-dpp})^{2+} < Os(biq)_2(\mu\text{-}2,5\text{-dpp})^{2+} \leq$$
$$Os(biq)_2(\mu\text{-}2,3\text{-dpp})^{2+} < Os(\mu\text{-}2,5\text{-dpp})_3^{2+} < Os(\mu\text{-}2,3\text{-dpp})_3^{2+} < Ru(bpy)_2(\mu\text{-}2,5\text{-}$$
$$dpp)^{2+} \leq Ru(bpy)_2(\mu\text{-}2,3\text{-dpp})^{2+} < Ru(biq)_2(\mu\text{-}2,5\text{-dpp})^{2+} \leq Ru(biq)_2(\mu\text{-}2,3\text{-}$$
$$dpp)^{2+} < Ru(bpy)(\mu\text{-}2,5\text{-dpp})_2^{2+} \leq Ru(bpy)(\mu\text{-}2,3\text{-dpp})_2^{2+} < Ru(m\text{-}2,3\text{-dpp})_3^{2+}.$$

In cyclometallated compounds, the oxidation processes are displaced to less positive potentials because of the presence of the anionic ligands. Such processes are usually irreversible presumably because they concern metal-based orbitals involved in bonding interactions.[44]

Polynuclear Compounds

In the polynuclear compounds the redox processes featured by the isolated component units are maintained, although some changes in the potential values may occur. The extent of these changes depends on the communication ability of the bridging ligand. If the bridging ligand does not allow substantial electronic communication, each component unit maintains its intrinsic redox properties. When the bridging ligand allows substantial electronic communication, only an estimate of the redox properties of a polynuclear complex can be made from the properties of the parent mononuclear complexes.[39]

It is well known[63] that supramolecular species containing a number of identical noninteracting centers exhibit current–potential responses having the same shape as that obtained with the corresponding molecule containing a single center. Only the magnitude of the current is enhanced by the presence of additional electroactive centers. Therefore in our compounds, equivalent, noninteracting units undergo electrochemical processes at the same potential. This allows us to control the number of electrons lost at certain potential by placing in the dendrimer the desired number of suitable, equivalent, and noninteracting units.

Oxidation Processes.   Because of the presence of several metal ions, each capable of undergoing an oxidation process, our polynuclear compounds show complex, very interesting oxidation patterns. For tetranuclear complexes like **4A**[18,30] and **4B**[18] (Scheme 1 and Table 1) a 3–1 oxidation pattern is expected, i.e., a three-electron process related to the oxidation at the same potential of the three peripheral, equivalent, and noninteracting $RuL_2(\mu\text{-}2,3\text{-dpp})^{2+}$ units followed by a one-electron process related to the oxidation of the central $Ru(\mu\text{-}2,3\text{-dpp})_3^{2+}$ component. The experimental results, however, support only the first process (Table 2) because the presence of the three contiguous, already oxidized peripheral components displaces the oxidation of the central metal ion at potentials more positive, practically outside the accessible potential window.

A 1–3 oxidation pattern (i.e., a one-electron process followed by a three-electron one) is exhibited by complex **4C**[17] (Scheme 1 and Table 1): the central, Os-containing unit is oxidized at +1.25 V, followed by simultaneous oxidation of the three peripheral, equivalent, and noninteracting $Ru(bpy)_2(\mu\text{-}2,3\text{-dpp})^{2+}$ components at +1.55 V (Table 2).

A different oxidation pattern (namely, a 1–2 one) is shown by the tetranuclear complex **4I**[64] (Scheme 1 and Table 1): the peripheral Os-containing unit is oxidized at +1.00 V, followed by simultaneous oxidation of the two peripheral Ru-containing units at +1.50 V (Table 2). The oxidation of the central $Ru(\mu\text{-}2,3\text{-dpp})_3^{2+}$ unit is displaced outside the accessible potential window by the presence of the three already oxidized units.

Table 2. Electrochemical Data in Acetonitrile versus SCE[*]

Compound	E_{ox}, V [n](site)	E_{red}, V [n](site)	Ref.
2A	+1.38 [1] (Ru), +1.55 [1] (Ru)	−0.67 [1] (BL), −1.17 [1] (BL), −1.57 [2] (bpy), −1.89 [2] (bpy)	18
2B	+1.57 [2] (Ru)	−0.45 [1] (BL), −0.81 [2] (biq), −0.95 [1] (BL), −1.19 [2] (biq)	18
2C	+1.36 [1] (Ru), +1.48 [1] (Ru)	−0.68 [1] (BL), −1.18 [1], −1.57 [2], −1.81 [2]	18
2D	+1.37 [1] (Ru), +1.54 [1] (Ru)	−0.53 [1] (BL), −1.08 [1] (BL), −1.50 [2] (bpy), −1.81 [2] (bpy)	18
2E	+1.48 [2] (Ru)	−0.45 [1] (BL), −0.82 [2] (biq), −0.99 [1] (BL), −1.26 [2] (biq)	18
2F	+1.43 [2] (Ru)	−0.47 [1] (BL), overlapping waves	18
2G	+0.99 [1] (Os), +1.52 [1] (Ru)	−0.65 [1] (BL), −1.11 [1] (BL), overlapping waves	57
2J	+0.90 [1] (Os), +1.20 [1] (Os)	−0.68 [1] (BL), −1.10 [1] (BL), −1.38 [2] (bpy), −1.62 [2] (bpy)	47c
2K	+0.92 [1] (Os), +1.22 [1] (Os)	−0.56 [1] (BL), −1.00 [1] (BL), −1.46 [2] (bpy), −1.71 [2] (bpy)	47c
2L	+1.82 [1] (Ru)	−0.64 i (BL), −0.78 i [2] (Medpp$^+$), −0.92 i (Medpp$^+$), −1.03 i (Medpp$^+$), −1.12, −1.21, −1.46 [multielectronic]	47d
3A	+1.48 [2] (Ru$_p$)[†]	−0.71 two overlapping waves	47b
	+1.48 [2] (Ru$_p$)[†]	−0.55 [1] (BL), −0.75 [1] (BL), −1.17 [1] (BL), −1.47 [2] (bpy+BL), −1.75 [2] (bpy)	18
3B	+1.62 [2] (Ru$_p$)[†]	−0.63 two overlapping waves	47b
	+1.60 [2] (Ru$_p$)[†]	−0.47 [2] (BL), −0.87 [2] (biq), −1.17 [2] (BL), −1.50 [1] (bpy), −1.79 [2] (biq)	18
3C	+1.45 [2] (Ru$_p$)[†]	−0.66 two overlapping waves	47b
	+1.45 [2] (Ru$_p$)[†]	−0.48 [1] (BL), −0.60 [1] (BL), −1.10 [1] (BL), −1.30 [1] (BL), −1.52 [1] (bpy)	18
3D	+1.60 [2] (Ru$_p$)[†]	−0.55 two overlapping waves	47b
	+1.57 [2] (Ru$_p$)[†]	−0.47 [2] (BL), −0.89 [2] (biq), −1.21 [2] (BL), −1.53 [1] (bpy), −1.78 [1] (biq)	18
3E			
3F			
3G	+0.72 [1] (Ru), +1.45 [2] (Ru)		27
3H	+0.88 [1] (Ru), +1.63 [2] (Ru)		27
3I			
3J	+0.75 [1] (Ru), +1.15 [2] (Os)		27
4A	+1.50 [3] (Ru$_p$)[†]	−0.56 [1] (BL), −0.63 [1] (BL), −0.70 [1] (BL), −1.20 [1], −1.33 [1], −1.48 [1]	18
	+1.53 [3] (Ru$_p$)[†]	−0.62 [1] (BL), −0.77 [1] (BL), −1.23 [1] (BL)	30
4B	+1.58 [3] (Ru$_p$)[†]	−0.6 [3] (BL), −0.87 [3] (biq), −1.15 [3] (BL)	18
4C	+1.25 [1] (Os), +1.55 [3] (Ru)	−0.55 [1] (BL), −0.65 [1] (BL), −0.77 [1] (BL)	17
4D	+1.40 [3] (Ru)	−0.58, −0.65, −0.80, −1.10, −1.30	57
4E	+1.67 [3] (Ru)	−0.52, −0.87, −1.10, −1.60, −2.00	57

(continued)

Table 2. (Continued)

Com- pound	E_{ox}, V [n](site)	E_{red}, V [n](site)	Ref.
4F	+0.92, +1.10, +1.37	−0.64, −0.80, −0.92, −1.22, −1.38	57
4I	+1.00 [1] (Os), +1.50 [2] (Ru)		64
4J	+1.82 [1] (Ru$_i$)†	−0.79 [6] (2,3-Medpp$^+$), −0.98 [3] (BL)	30
4L	adsorption	adsorption	30
4M§	+1.25 i [~1] (Ru)	−0.48 [1] (BL), −0.60 [1] (BL), −0.70 [1] (BL), −1.19 [1] (BL), −1.37 [1] (BL), ≈−1.60	21
4N§	+1.40 i [>3] (Ru+Ir)	−0.44 [1] (BL), −0.55 [1] (BL), −0.65 [1] (BL), −1.12 [1] (BL), −1.29 [1] (BL), ≈−1.53	21
4O§	+0.75 i [1] (Os), +1.50 i [>1] (Rh)	−0.35 i [1] (BL), −0.54 i [1] (BL), −0.82 i [1] (BL)	21
4P§	+1.00 [1] (Os), +1.40 i [3] (Ir)	−0.32 [1] (BL), −0.49 [1] (BL), −0.67 [1] (BL), −1.11 [1] (BL), −1.29 [1] (BL), ≈−1.44, −1.60	21
6A	+1.44 [4] (Ru$_p$)†	−0.55 overlapping waves	22,23
6B	+1.38 [3] (Ru$_p$)†	−0.50 overlapping waves	22, 23
6C	+1.41 [4] (Ru$_p$)†	−0.50 overlapping waves	22, 23
6D	+1.36 [4] (Ru$_p$)†	−0.50 overlapping waves	22, 23
6E	~+1.8 [4] (Ru$_p$)†		23
6F	+1.06 [4] (Os)		23
6G	+1.50 [2] (Ru-bpy), +1.66 [2] (Ru-biq)		23
6H	+1.05 [2] (Os), +1.45 [2] (Ru$_p$)†		23
6I	+1.31 [4] (Ru)	−1.10 [2] (BL), −1.21 [1] (BL), −1.31 [1] (BL), −1.48 [4] (bpy), −1.74	24
7A	+1.38 [3] (Ru$_p$)†	−0.58 [3] (BL)	25
10A	+1.43 [6] (Ru$_p$)†		26
	+1.53 [6] (Ru$_p$)†	−0.73 [6] (BL$_o$), −1.22 [3] (BL$_i$)‡	30
10B	+1.62 [6] (Ru$_p$)†		27
10C	+1.17 [1] (Os), +1.50 [6] (Ru$_p$)†		27
10D	+1.24 [1] (Os), +1.59 [6] (Ru$_p$)†		27
10E	+1.05 [6] (Os$_p$); +1.39 [1] (Os$_c$)†		27
10F	+1.00 [6] (Os$_p$)†		27
10G	+1.83 [3] (Ru$_i$)†	−0.75 [12] (2,3-Medpp$^+$), −0.95 [3] (BL$_i$)‡	30
10H	~+1.69 [~6] adsorption	adsorption	30
13A	+1.50 [9] (Ru$_p$)†		28
22A	+1.52 [12](Ru$_p$)†	overlapping waves	29, 30

Notes: *[n] is the number of exchanged electrons; (X) is the site involved in the redox process.
§In CH_2Cl_2.
†M$_c$, M$_i$, M$_p$ stand for central, intermediate, and peripheral metal, respectively.
‡BL$_i$ and BL$_o$ stand for inner and outer bridging ligand, respectively.

A 4–1–1 pattern is expected for the six "symmetrical" hexanuclear compounds **6A–F**[22,23] (Scheme 1 and Table 1), i.e., a first four-electron process attributable to the oxidation of the four equivalent, noninteracting peripheral metal-containing units and, at more positive potential, two successive one-electron processes corresponding to the oxidation of the two interacting inner Ru-containing units. The differential pulse voltammetry, however, shows only a peak which, on the basis of the potential value (Table 2) and the number of the involved electrons, can be assigned to the oxidation of the peripheral metal-containing units. The oxidation of the two inner Ru-containing units is not observed because it is displaced outside the accessible potential window by the presence of the already oxidized peripheral units. Figure 11 shows the differential pulse voltammetry results obtained for **6D**

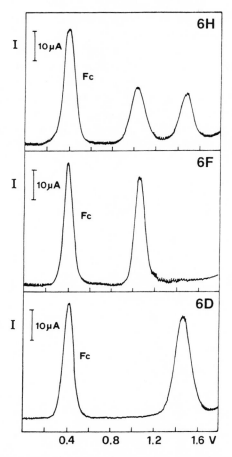

Figure 11. Oxidation pattern for some hexanuclear complexes. Fc indicates the oxidation peak of ferrocene, used as an internal standard.

and **6F**. For the former the peak concerns the simultaneous oxidation of the four peripheral $Ru(bpy)_2(\mu\text{-2,5-dpp})^{2+}$ units, whereas for the latter that of the $Os(bpy)_2(\mu\text{-2,3-dpp})^{2+}$ units.

For compound **6H**[23] (Scheme 1 and Table 1) the oxidation pattern is quite different; the differential pulse voltammetry exhibits two peaks of equal height, both corresponding to a two-electron oxidation process (Figure 11). The first oxidation occurs at nearly the same potential as the four-electron process of compound **6F**. This shows that, as expected, the two $Os(bpy)_2(\mu\text{-2,3-dpp})^{2+}$ units are the first to be oxidized (Table 2). The second process concerns the oxidation of the two $Ru(bpy)_2(\mu\text{-2,5-dpp})^{2+}$ units. Since such units lie far away from the previously oxidized Os-containing units, their oxidation occurs at a potential (Table 2) close to that of the equivalent peripheral units of **6D**. As in the case of the compounds **6A–F** the oxidation of the two inner units are displaced outside the accessible potential window.

Very interesting oxidation patterns are obtained for decanuclear compounds (Scheme 1 and Table 1). For example, in compound **10C**,[27] which contains one Os^{2+} and nine Ru^{2+} ions, the Os^{2+} ion is expected to be oxidized at less positive potentials than the nine Ru^{2+} ions. Furthermore, because of the different electron donor properties of the ligands, the six peripheral Ru^{2+} ions are expected to be oxidized at less positive potentials than the three intermediate Ru^{2+} ions. In agreement with these expectations, the differential pulse voltammogram of **10C** (Figure 12) shows

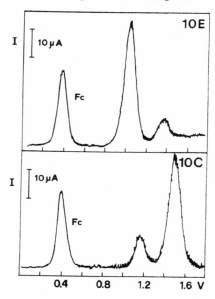

Figure 12. Oxidation pattern for some decanuclear complexes. Fc indicates the oxidation peak of ferrocene, used as an internal standard.

an oxidation peak at +1.17 V which, on the basis of the potential value can be assigned to the one-electron oxidation of the central Os^{2+} metal ion, and another peak at +1.50 V (Table 2), which has the same bandwidth but a six times higher intensity. This peak can be assigned to the independent one-electron oxidation of the six peripheral noninteracting Ru^{2+} ions. Oxidation of the three intermediate Ru^{2+} ions is further shifted toward more positive potentials and cannot be observed in the accessible potential window. For **10E**[27] one expects that oxidation involves first the six peripheral Os^{2+} ions (which contain the stronger electron donor bpy ligand in their coordination sphere), and then the central one. This is fully consistent with the differential pulse voltammetry results (Figure 12). The more positive potential for the oxidation of the central Os^{2+} ion (Table 2) in **10E** compared to that of the same metal ion in **10C** is obviously due to the presence of six already oxidized osmium ions when the central ion of **10E** undergoes oxidation. Following the same consideration it is easy to understand why the oxidation of the intermediate Ru^{2+} units is not observed.

Interesting information regarding the influence of positive-charged terminal ligands can be obtained by comparing the electrochemical behavior of **10A** and **10G**.[30]

For **10A** the differential pulse voltammetry exhibits only one six-electron peak which corresponds to the simultaneous one-electron oxidation of the six peripheral, noninteracting Ru^{2+} units (Table 2 and Figure 13). Oxidation of the central and intermediate metal ions cannot be observed in the accessible potential window.

For compound **10G** the differential pulse voltammetry (Figure 13) shows again only one oxidation peak, but in this case the involved electrons are three. The number of the electrons and the potential value (Table 2), more positive than in the case of **10A**, suggest that the oxidation involves the three equivalent intermediate Ru^{2+} units. As one can see from Figure 13, the peak is somewhat broader than that observed for **10A**, indicating a certain degree of interaction among the units involved in this process. The oxidation of the central Ru^{2+} unit is displaced outside the accessible potential window because of the interaction with the nearby oxidized intermediate units. Even the oxidation of the six peripheral units is not observed because it is displaced to more positive potentials by the presence of the positive charge on the 2,3-Medpp$^+$ terminal ligands, as indicated by very high potential for the oxidation of $Ru(2,3\text{-Medpp})_3^{5+}$.[47d]

The increase of the potential, at which the oxidation of the metal ions coordinated to the 2,3-Medpp$^+$ ligand takes place, is further evidenced by the different behavior of the tetranuclear compound **4A** and **4J**[30] (Scheme 1 and Table 1). As clearly indicated by the differential pulse voltammetry, which exhibits a three-electron oxidation peak for the former and a one-electron oxidation peak for the latter (Figure 14), in **4A** the oxidation involves the three peripheral Ru^{2+} units (as mentioned before), whereas in **4J**, only the central Ru^{2+} unit can be oxidized in the accessible potential window (Table 2).

In agreement with the behavior shown by **10A**, the oxidation of **22A**[29,30] (Scheme 1 and Tables 1 and 2) involves the peripheral metal ions which are equivalent, as

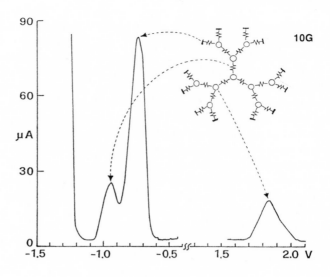

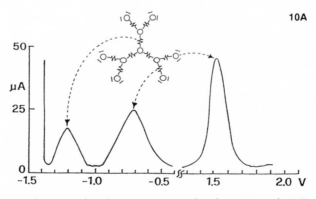

Figure 13. Oxidation and reduction patterns for the **10A** and **10G** decanuclear complexes.

indicated by the presence of a twelve-electron oxidation peak in the differential pulse voltammetry. As for **10A**, no successive oxidation can be observed, presumably because of the large positive charge accumulated in the periphery.

It is interesting to note that for **4A**, **10A**, and **22A** the oxidation, which involves the same peripheral $Ru(bpy)_2(\mu\text{-}2,3\text{-}dpp)^{2+}$ units, takes place practically at the same potential (Table 2). This shows that the electrochemical properties of a certain unit depend only on the neighboring units, regardless of the nuclearity and the overall electric charge of the supramolecular structure.

The hexanuclear cyclometallated complex $6I^{24}$ (Scheme 1 and Table 1) shows a single reversible oxidation process at +1.31 V (Table 2), which involves four electrons. This process can be attributed to the oxidation of the four, almost noninteracting, peripheral Ru(II) metals. The lack of other oxidation processes is an expected result since the increase (+4) in the charge of the complex displaces the oxidation of the core outside the accessible potential window.

Reduction Processes.   Because of the presence of many polypyridine ligands, each capable of undergoing several reduction processes,[62] the electrochemical reduction of the polynuclear compounds produces very complex patterns.

As far as the tetranuclear complexes (Scheme 1 and Table 1) are concerned, the differential pulse voltammetry of compound $4A^{30}$ exhibits three one-electron reduction peaks starting at −0.62 V (Table 2), which can be attributed to successive reduction of interacting bridging ligands (Figure 14). For potential ≤ −1.4 V other overlapping waves are present (not shown in Figure 14) that can be attributed to the second reduction of the bridging ligands followed by the reduction of the bpy ligands. The differential pulse voltammetry of compound $4J^{30}$ shows two peaks (Table 2): a six-electron peak at −0.79 V that can be assigned to the one-electron reduction of the six methylated peripheral ligands at nearly the same potential and, at more negative potential, a broad peak involving three electrons that can be assigned to the reduction of the three bridging ligands (Figure 14). For potential ≤ −1.2 V unresolved peaks are observed (not shown in Figure 14) that could represent the second reduction of the methylated ligands followed by the second reduction of the bridging ligands.

Compound $10A^{30}$ (Scheme 1 and Table 1) shows a differential pulse voltammetry (Figure 13) with two, broad peaks (at −0.73 V and −1.22 V, Table 2) followed by several other overlapping peaks. The first peak, which corresponds to six electrons, may be attributed to the one-electron reduction of the six outer equivalent bridging ligands. Since the six processes take place at approximately the same potential, such six ligands interact only slightly with one another. This would suggest that the LUMO orbital mainly involves the more external chelating site of the 2,3-dpp bridging ligand. The second peak, which involves three electrons, should represent the one-electron reduction of the three inner bridging ligands, again mainly on their external chelating site. The overlapping peaks present in the voltammograms at more negative potentials (not shown in Figure 13) should concern the second reduction of the bridging ligands followed by the reduction of the bpy ligands.

In the differential pulse voltammetry (Figure 13) of compound $10G^{30}$ (Scheme 1 and Table 1), two peaks are present (−0.75 and −0.94 V, Table 2), followed by several other overlapping peaks. The first peak, which involves 12 electrons, corresponds to the one-electron reduction of the 12 outer methylated ligands. The second peak, which involves three electrons, should concern the one-electron reduction of the three inner bridging ligands. The six intermediate bridging ligands, because of their closeness to the already reduced outer ligands, can be reduced only

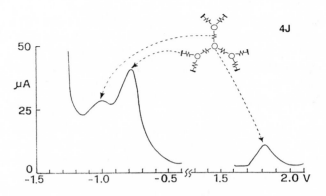

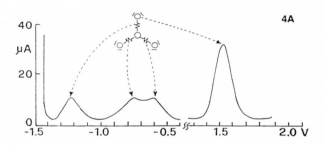

Figure 14. Oxidation and reduction patterns for the **4A** and **4J** complexes.

at more negative potential in comparison with the three inner ligands. This is also in agreement with the fact that in **10G** the reduction of the inner bridging ligands takes place at a potential less negative than in the case of **10A** where the inner ligands are close to the already reduced six intermediate ligands. The unresolved peaks observed at potential ≤ -1.1 V (not shown in Figure 13) should concern the reduction of the six intermediate bridging ligands overlapping the second reduction of the methylated ligands.

For **22A**,[30] both the cyclic and differential pulse voltammetry patterns show the presence of many overlapping waves that can hardly be assigned.

It is instructive to see how the reduction potentials of the three 2,3-dpp ligands of the central core change in going from the mononuclear $Ru(2,3\text{-dpp})_3^{2+}$ to the tetra- and decanuclear complexes (Figure 15).[30]

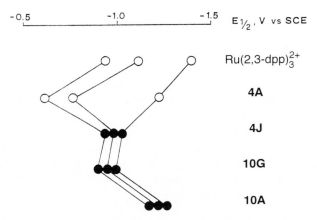

Figure 15. Correlation between the reduction potentials of the core 2,3-dpp ligands in various complexes. The black circles indicate that the reduction of the core ligand occurs after reduction of other sites (see text).

In $Ru(2,3\text{-dpp})_3^{2+}$ the three ligands are reduced at different potentials, showing that there is a noticeable ligand–ligand interaction through the central metal. In **4A** the three processes are still well separated, but they occur at less negative potentials because the LUMO of each bridging ligand is stabilized by $[Ru(bpy)_2]^{2+}$ coordination. For **4J**, **10G**, and **10A**, reduction of the central bridging ligands occurs after reduction of other sites (Table 2, Figures 13 and 14). Under such conditions, the three reduction waves almost coalesce, indicating that the ligand–ligand interaction through the central metal decreases. A determinant factor for such through-metal interaction is the energy difference between the π^*-LUMO of the ligands and the $\pi\text{-}t_{2g}$ orbitals of the metal. The decrease in the ligand–ligand interaction could then be explained by considering that an increase in the LUMO energy of the central bridging ligands caused by the reduction of external units increases the energy difference between ligand LUMO and metal orbitals.[30]

As shown in Figure 15, reduction of the central bridging ligand in **4J** occurs at more negative potentials than in **10G** because the already reduced sites are closer in the former compound.[30] At first sight it may appear strange that the reduction of the central bridging ligands of **10A** occurs at approximately the same potential as the third reduction of **4A**, in spite of the presence of six outer reduced sites in the former compound. One should consider, however, that in **4A** the third reduction is displaced to negative potentials by the ligand–ligand interaction, which is strongly reduced in **10A** (see above).

The hexanuclear cyclometallated compound $6I^{24}$ (Scheme 1 and Table 1) shows five reduction processes (Table 2). The first one, which occurs at a potential very close to the first reduction of the model compound $[(bpy)_2Ru(dppnH)]^{2+}$, is reversible and involves two electrons. It can be assigned to the simultaneous one-electron reduction of two remote bridging ligands. More specifically, each added electron probably enters a LUMO orbital largely localized on the N–N moiety (Ru bonded) of the bridging ligands. The two successive processes are monoelectronic and can be assigned to reduction of the third and fourth bridging ligands (in their N–N moieties) for the following reasons: (1) as shown by the behavior of model compounds, reduction of the bpy ligands, and even more that of the C^-–N moiety of the bridging ligands, are expected to occur at more negative potentials; (2) since **6I** contains two series of four equivalent bpy ligands, the first reduction involving bpy ligands should be a four-electron process. The displacement of the reduction of the third bridging ligand at a more negative potential compared to the (simultaneous) reduction of the first and second ones is due to the presence of an already reduced bridging ligand on the same Rh center. The further displacement of the reduction of the fourth bridging ligand is due to the enhanced interaction across the core with increasing number of added electrons. A similar behavior has been observed for the reduction of the peripheral ligands of dinuclear metal complexes.[47c,65] The fourth reduction process can be assigned to reduction of four equivalent bpy ligands on the basis of the number of electrons involved and potential value, which is almost the same as that of the bpy reduction in the model compound. Experimental problems due to the adsorption of the compound on the electrode prevent to establish the number of the electron involved in the fifth reduction process. It is, however, possible an estimate of the potential value that agrees with that expected for first reduction of the second series of the four bpy ligands.

In conclusion, the electrochemical data offer a fingerprint of the chemical and topological structure of the polynuclear compounds. Furthermore, made-to-order synthetic control of the number of electrons exchanged at a certain potential can be achieved. The presence of multielectron processes makes such polynuclear complexes very attractive in view of their possible application as multielectron-transfer catalysts.[66] Examination over a more extended oxidation potential window (in a solvent like liquid SO_2) should permit one to obtain an even larger variety of oxidation patterns.

C. Absorption Spectra

When only polypyridine-type ligands are present, each mononuclear metal-based unit exhibits intense LC bands in the UV region and moderately intense MLCT bands in the visible. As it is shown by the electrochemical behavior (*vide supra*), in the polymetallic species there is some interaction among the neighboring metal-based units. To a first approximation, however, each building block carries its own absorption properties in the polynuclear species so that the molar absorption coefficients exhibited by the compounds of higher nuclearity are huge, as it is clear by a cursory examination of the absorption data reported in Table 3. For example, the spectra of the decanuclear compounds **10B** and **10C**[27] (Scheme 1 and Table 1)

Table 3. Photophysical Data

Compound	Solvent	Abs (nm)	ε $(M^{-1}cm^{-1})$	Em_{RT} (nm)	τ_{RT} (ns)	Φ_{RT}	Em_{77K} (nm)	τ_{77K} (μs)	Ref.
2A	AN	527	24200	802 c	102 a, 125 d	3×10^{-3}d			18
	Et/Met						709 c	2.00	18
2B	AN	537	15600	789 c	65 d				18
	Et/Met						720 c	1.73	18
2C	AN	543	22400	799 c	75 d				18
	Et/Met						723 c	1.55	18
2D	AN	585	15900	824 c	155 a				18
	Et/Met						771 c	0.94	18
2E	AN	609	11500	820 c	170 a				18
	Et/Met						722 c	1.82	18
2F	AN	595	13700	830 c	190 a				18
	Et/Met						792 c	1.00	18
2G	AN	543	23500	>850					57
2H	AN	601	19200						59
	Et/Met						968 c[†]		59
2I	AN	620	17900						59
	Et/Met						900 c[†]		59
2J	AN	555	21800						47c
	Et/Met						928 c[†]		59
2K	AN	616	27000						47c
	Et/Met						984 c[†]		59
2L	AN	457	19900	668 c	870 d	3.3×10^{-2}d			47d
	Et/Met						607 c	5.5	47d
3A	AN	545	23500	766 u	75 a				47b
	AN			804 c	80 d	1×10^{-3}d			18
	Et/Met						712 u	1.78	47b
	Et/Met						721 c		18
3B	AN	546	28700	742 u	142 a				47b
	AN			773 c					18
	Et/Met						702 u	2.18	47b
	Et/Met						713 c		18
3C	AN	592	28100	814 u	64 a				47b
	AN			831 c					18
	Et/Met						755 u	0.98	47b
	Et/Met						767 c		18
3D	AN	591	22900	774 u	120 a				47b
	AN			805 c	190 d	6×10^{-3}d			18
	Et/Met						734 u	1.28	47b
	Et/Met						739 c		18
3E	AN	559	36300						59
	Et/Met						896 c[†]		59

(continued)

Table 3. (Continued)

Com-pound	Solvent	Abs (nm)	ε $(M^{-1}cm^{-1})$	Em_{RT} (nm)	τ_{RT} (ns)	Φ_{RT}	Em_{77K} (nm)	τ_{77K} (μs)	Ref.
3F	AN	606	33800						59
	Et/Met						980 c[†]		59
3G	AN	618	24300	>880			>880		27
3H	AN	624	24000	>880			>880		27
3I									
3J	AN	600	25200	>880			>880		27
4A	AN	545	46000	811 c	50 a, 60 d	1×10^{-3}d			18
	Et/Met						727 c	1.38	18
4B	AN	610	41500	795 c	130 a, 190 d	1×10^{-3}d			18
	Et/Met						725 c	1.86	18
4C	AN	549	40000	875 c	18 d				17
	Et/Met						802 c	0.41	17
4D	AN	649	25400	810 u	158 d				57
4E	AN	700	30500	800 u	100 d				57
4F	AN	618	24600						59
	Et/Met						960 c[†]		59
4G	AN	552	41700	812 u	44 a				19
	Et/Met						752 u	0.90	19
4H	AN			768 u	110 a				20
	Et/Met						720 u	1.5	20
4I	AN	547	37500	>880					64
4J	AN	505sh	38000	714 c	600 d	6.5×10^{-3}d			30
	Et/Met						698 c	4.3	30
4K	AN	526		812 u	445 a				16
4L	AN	461	38200	722 c	300 d	1.8×10^{-4}d			30
	Et/Met						698 c	3.6	30
4M	CH_2Cl_2	471	15400	681 c	330 d	1.8×10^{-2}d	645 c	1.05, 3.40	21
4N	CH_2Cl_2	499	31500	812 c	2.2 d	1.5×10^{-4}d	726 c	0.57, 1.71	21
4O	CH_2Cl_2	515	23300	821 c	55 d	5.8×10^{-3}d	789 c	0.62	21
4P	CH_2Cl_2	534	25400	825 c	5.2 d	4.5×10^{-4}d	810 c	0.43	21
6A	AN	540	59000	770 u	53 a				22
	Et/Met						716 u	1.33	22
6B	AN	582	52000	810 u	40 a				22
	Et/Met						756 u	0.83	22
6C	AN	540	62200	768 u	55 a				22
	Et/Met						716 u	1.30	22
6D	AN	577	54100	812 u	44 a				22
	Et/Met						764 u	0.83	22

(continued)

Table 3. (Continued)

Com-pound	Solvent	Abs (nm)	ε ($M^{-1}cm^{-1}$)	Em_{RT} (nm)	τ_{RT} (ns)	Φ_{RT}	Em_{77K} (nm)	τ_{77K} (μs)	Ref.
6E	AN	571	60200	760 u	80 a				23
	Et/Met						716 u	1.5	23
6F	AN	560	81500						23
	Et/Met						912 c[†]		59
6G	AN	588	39000	790 u	41 a				23
	Et/Met						750 u	1.27	23
6H	AN	576	50300	802 u	39 a				23
	Et/Met						752 u	0.82	23
6I	[CH$_2$Cl]$_2$	450	53300	675 c	925 d	0.038 d			24
	[CH$_2$Cl]$_2$/CH$_2$Cl$_2$						632 c	4.5	24
7A	AN	547	76200	80	80d	9×10^{-4}d			25
	Et/Met						725	2.0	25
10A	AN	541	125000	809 c	55 d	1×10^{-3}d			26
	Et/Met						732 c	1.3	26
10B	AN	555	109500	789 c	130 d	6×10^{-3}d			27
	Et/Met						722 c	1.65	27
10C	AN	550	117000	808 c, 860 c	65 d	5×10^{-4}d			27
	Et/Met						720 c	1.33	27
10D	AN	556	117500	789 c, 860 c	125 d	3×10^{-4}d			27
	Et/Met						722 c	1.71	27
10E	AN	563	140500						27
	Et/Met						900 c		59
10F	AN	560	132500						27
	Et/Met						892 c		59
10G	AN	550sh	40000	668 c	600 d	3.9×10^{-4}d			30
	Et/Met						649 c	5.1	30
10H	AN	500sh	30000	750 c	too weak	$<3\times10^{-5}$d			30
	Et/Met						too weak	too weak	30
13A	AN	544	133000	800 c	62 d				28
	Et/Met						722 c	1.46	28
22A	AN	542	202000	786 c	45 d	3.0×10^{-4}d			29, 30
	Et/Met						730 c	1.4	29, 30

Note: [†]At 90 K.

display absorption bands with ε up to 600 000 M^{-1} cm^{-1} in the UV region and up to 140 000 M^{-1} cm^{-1} in the visible region (Figure 16). The bands with maxima at 262 and 380 nm can be assigned to $\pi \to \pi^*$ transitions of the biq ligands, the band at 282 nm to $\pi \to \pi^*$ transitions on the bpy ligands, and the broad absorption in the 300–350 nm region to $\pi \to \pi^*$ transitions on the biq and 2,3-dpp ligands. The broad bands observed in the visible region receive contribution from several types of MLCT transitions. The energies of these transitions depend on the nature of the donor metal ion, the acceptor ligand, and the ancillary ligands. Even in the case of the homometallic complex **10B**, five different types of "proximate"[67] MLCT transitions are expected because of the presence of three nonequivalent positions for the metals (M_c, M_i, and M_p) and two nonequivalent positions for the bridging ligands (see Figure 4).

From the spectroscopic and electrochemical results, the following trends have been established for the energy ordering of the metal-to-ligand CT transitions in polynuclear complexes of 2,3- and 2,5-dpp bridging ligands (BL): (1) for the same acceptor ligand and the same ancillary ligands, Os$^{2+} \to$acceptor ligand < Ru$^{2+} \to$acceptor ligand; (2) for the same metal and ancillary ligands, M$\to$BL < M$\to$biq <

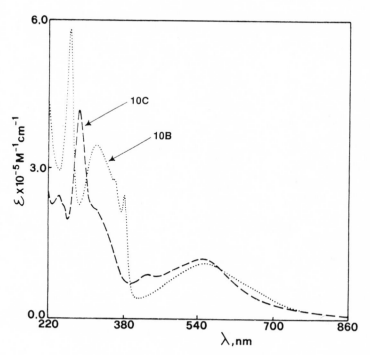

Figure 16. Absorption spectra of compounds **10B** and **10C** in acetonitrile solution at room temperature.

M→bpy; and (3) for the same metal and the same acceptor ligands, $(bpy)_2M$→acceptor ligand < $(biq)_2M$→acceptor ligand < $(BL)_2M$→acceptor ligand. From a comparison of the absorption spectra of **10B** and **10C** (Figure 16) it is easy to assign the absorption band at about 430 nm to M→bpy CT transitions. The broad band with maximum at ~550 nm receives contributions from M→biq and M→bridging ligand CT transitions. It should be noted that in the Os-containing compounds the absorption bands extend to the red because the probability of the singlet→triplet transitions is relatively high owing to the enhanced spin-orbit coupling. The above discussion of the absorption spectra of the decanuclear complexes can straightforwardly be extended to the other complexes of the same family.

In the "protected" dendrimers, the presence of methylated peripheral ligands causes strong modifications in the absorption spectra.[30] The peripheral Ru^{2+} ions of **10G** (Scheme 1 and Table 1) are more difficult to oxidize than the Ru^{2+} ions carrying bpy as peripheral ligands (Table 2); as a consequence, the energy of the (peripheral ligand)$_2Ru$→(μ-2,3-dpp) CT transition moves to the blue on passing from the "bpy" dendrimers to the "protected" ones. At the same time, the Ru→(2,3-Medpp$^+$) CT transition is expected to be at substantially lower energy than the corresponding Ru→bpy CT transition because of the different acceptor properties of bpy and 2,3-Medpp$^+$. The Ru→(peripheral ligand) and Ru→(bridging ligand) CT bands tend therefore to merge in going from **10A** (Scheme 1 and Table 1) to **10G**. Furthermore, "remote" CT transitions involving the peripheral ligands are

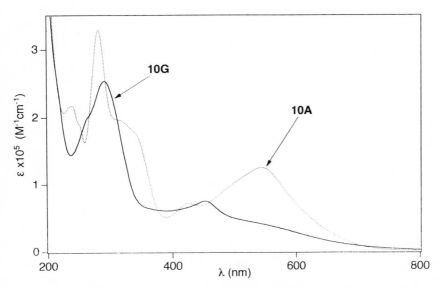

Figure 17. Absorption spectra of compounds **10A** and **10G** in acetonitrile solution at room temperature.

lowered in energy. The result of such a situation is a poorly resolved absorption spectrum for the "protected" dendrimers (Figure 17).

The absorption spectrum of the hexanuclear cyclometallated compound **6I**[24] (Scheme 1 and Table 1) is shown in Figure 18. This complex can be viewed as made of two types of (interacting) chromophores, the four peripheral Ru(II)-based polypyridine-type units, and the Rh(III)-based cyclometallated core (see Figure 5, bottom). The spectrum of the reference compound $[Ru(bpy)_2(dppnH)]^{2+}$ shows a very intense band with maximum at 286 nm which can be attributed to ligand-centered (bpy and dppnH) transitions, and a broad band with maximum at 433 nm which is due to MLCT transitions (Ru→bpy at higher energy and Ru→dppnH at lower energy, according to the reduction potentials).[24] The spectrum of the other reference compound, $[(ppy)_2Rh(\mu\text{-}Cl)_2Rh(ppy)_2]$, shows an intense, presumably LC, absorption band with maximum below 250 nm, and a series of weaker bands between 300 and 400 nm, largely of MLCT character.[24] It should be noted, however, that the spectrum of $[Ru(bpy)_2(dppnH)]^{2+}$ is more intense than the spectrum of $[(ppy)_2Rh(\mu\text{-}Cl)_2Rh(ppy)_2]$ throughout the near UV and visible spectral region. Since **6I** is made of four $[Ru(bpy)_2(dppnH)]^{2+}$-type units and only one $[(ppy)_2Rh(\mu\text{-}Cl)_2Rh(ppy)_2]$-type moiety, one can expect that, in the absence of strong intercomponent perturbations, the absorption spectrum of **6I** is dominated by the bands of

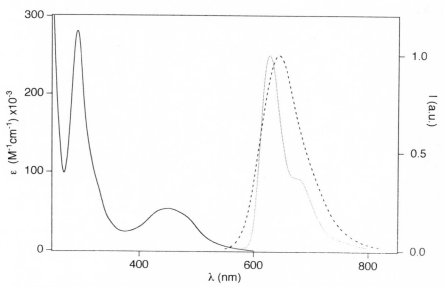

Figure 18. Absorption spectrum (*solid line*: 1,2-dichloroethane solution, room temperature) and uncorrected luminescence spectra (*dashed line*: 1,2-dichloroethane solution, room temperature. *dotted line*: 1,2-dichloroethane/dichloromethane 1:1 v/v rigid matrix, 77 K) of compound **6I**.

the $[Ru(bpy)_2(dppnH)]^{2+}$ units. This is in fact the case, but the spectrum of **6I** is slightly different from the sum of the spectra of its components, particularly in the visible region.

D. Luminescence Properties

As we have seen above, in complexes of polypyridine-type ligands, metal–metal interaction is weak for metals coordinated to the same bridging ligand, and almost negligible for metals that are far apart. Similarly, ligand–ligand interaction is appreciable only for ligands coordinated to the same metal. As a consequence, the energy levels of the component building blocks are essentially maintained. Light excitation in the visible absorption bands populates ^{1}MLCT excited states of the various components. Investigations carried out on $Ru(bpy)_3^{2+}$ with fast techniques indicate that the originally populated ^{1}MLCT excited states undergo relaxation to the lowest energy ^{3}MLCT level in the subpicosecond time scale.[68] If this behavior, as it seems likely, is of general validity for the various components of the polynuclear compounds, the actual result of light excitation is the population with unitary efficiency of the lowest energy ^{3}MLCT level of the component where light absorption has taken place. If each component were isolated, as it happens in mononuclear complexes, competition between radiative (luminescence) and radiationless decay to the ground state would account for the deactivation of the ^{3}MLCT level, with an overall rate constant, measured from the luminescence decay, in the range 10^6–10^8 s^{-1}. All the members of the mononuclear $M(BL)_{3-n}(L)_n^{2+}$ family, in fact, display a characteristic luminescence both in rigid matrix at 77 K and in fluid solution at room temperature. When the components are linked together in a supramolecular array, electronic energy can be transferred from an excited component to an unexcited one even if the electronic interaction is weak. In most of the examined polynuclear compounds, only one luminescence band, corresponding to the lowest energy ^{3}MLCT level, is observed (Table 3), indicating that energy transfer from upper-lying to lower-lying levels does occur (*vide infra*).

According to the above discussed energy ordering of the lowest excited state of the various building blocks, in **10A** (whose emission spectrum is shown in Figure 19) and **22A** compounds[30] (Scheme 1 and Table 1) the lowest energy excited state involves a Ru→(bridging ligand) CT transition in the peripheral units. In the **10G** "protected" dendrimer[30] (Scheme 1 and Table 1), the presence of methylated peripheral ligands causes strong modifications in the absorption (*vide supra*) and emission properties compared with those exhibited by the bpy analog **10A**. As we have seen above, the peripheral Ru^{2+} ions of **10G** are more difficult to oxidize than the Ru^{2+} ions carrying bpy as peripheral ligands (Table 2), and, as a consequence, the energy of the (peripheral ligand)$_2$Ru→(μ-2,3-dpp) CT transition moves to the blue on passing from the "bpy" dendrimers to the "protected" ones. Since excited states involving different units are expected to lie close in energy, the assignment of the luminescence of **10G** (Figure 19) to specific metal-based units is not

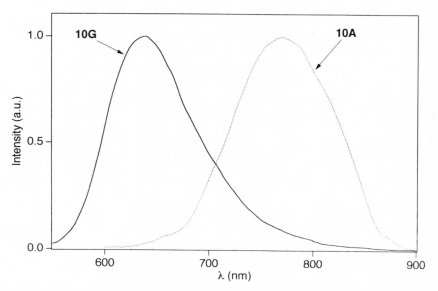

Figure 19. Luminescence spectra of compounds **10A** and **10G** in acetonitrile solution at room temperature.

straightforward. The electrochemical data suggest that the lowest MLCT excited state is a remote Ru(inner)→(2,3-Medpp$^+$) level. At present, we cannot say whether this level is involved in the luminescence process, but the small shift of the luminescence band in passing from fluid solution to rigid matrix would exclude this possibility.

The luminescence spectrum of the hexanuclear cyclometallated compound **6I**[24] (Scheme 1 and Table 1) in fluid solution at room temperature and in rigid matrix at 77 K (Figure 18) resembles that of its [Ru(bpy)$_2$(dppnH)]$^{2+}$ component and does not show any evidence of a [(ppy)$_2$Rh(μ-Cl)$_2$Rh(ppy)$_2$]-type emission. The luminescence lifetime and quantum yield (Table 3) are again similar, although not identical, to those found for [Ru(bpy)$_2$(dppnH)]$^{2+}$. The excitation spectrum ($\lambda_{em} =$ 650 nm) matches the absorption spectrum, even in the spectral region between 400 and 520 nm where the absorption spectrum of **6I** is significantly different from the absorption spectrum of its [Ru(bpy)$_2$(dppnH)]$^{2+}$ parent compound. These results indicate that: (1) the lowest excited state in the hexanuclear compound is substantially localized in the peripheral Ru-based units; (2) such an excited state involves the N–N chelating moiety of the bridging ligand, which is the ligand moiety easier to reduce; and (3) all the upper-lying excited states, including the lowest energy (potentially luminescent at 77 K) excited state of the core, are deactivated with unitary efficiency to the lowest excited state of the peripheral units.[24]

E. Intercomponent Energy Transfer: Antenna Effect

As we have seen above (Section IV.C), in the polynuclear complexes dealt with in this review it is possible to identify components which can undergo photoexcitation independently from one another. The excited component can then give rise to intercomponent energy transfer processes, in competition with intracomponent decay. For most of the components which constitute the examined systems, the lifetime of the lowest excited state is long enough to allow the occurrence of energy transfer to nearby components when suitable energetic and electronic conditions are satisfied. This is not usually the case for upper excited states, which usually decay very rapidly (picosecond time scale) to the lowest excited state within each component.

Electronic energy transfer in polynuclear metal complexes can be exploited for light harvesting (antenna effect). Generally speaking, an artificial antenna is a multicomponent system (Figure 20) in which several molecular components absorb the incident light and channel the excitation energy to a common acceptor component.[69]

Towards the development of artificial antenna systems based on metal complexes, one trend is to assemble very large numbers of molecular components in more or less statistical ways. For example, photon-harvesting polymers are being developed[70] in which rapid energy migration among pendant chromophoric groups is observed under certain conditions. Though very interesting, such systems, as well as the beautiful porphyrin-based antenna arrays,[71] are outside the scope of this

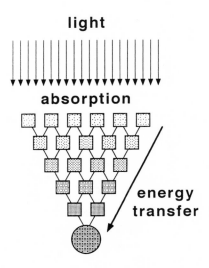

Figure 20. Schematic representation of an artificial antenna for light harvesting.

chapter and will not be discussed. We will only recall a few examples concerning the metal-based dendrimers.

Following the modular synthetic strategy described above, polynuclear species can be obtained with a high degree of synthetic control in terms of the nature and position of metal centers, bridging ligands, and terminal ligands. The energy of the lowest MLCT excited state of each unit depends on metal and ligands in a well-known and predictable way:

$$Os(bpy)_2(\mu\text{-}2,5\text{-}dpp)^{2+} \leq Os(bpy)_2(\mu\text{-}2,3\text{-}dpp)^{2+} < Os(biq)_2(\mu\text{-}2,5\text{-}dpp)^{2+} \leq$$

$$Os(biq)_2(\mu\text{-}2,3\text{-}dpp)^{2+} < Os(\mu\text{-}2,5\text{-}dpp)_3^{2+} < Os(\mu\text{-}2,3\text{-}dpp)_3^{2+} < Ru(bpy)_2(\mu\text{-}2,5\text{-}$$

$$dpp)^{2+} \leq Ru(bpy)_2(\mu\text{-}2,3\text{-}dpp)^{2+} < Ru(biq)_2(\mu\text{-}2,5\text{-}dpp)^{2+} \leq Ru(biq)_2(\mu\text{-}2,3\text{-}$$

$$dpp)^{2+} < Ru(bpy)(\mu\text{-}2,5\text{-}dpp)_2^{2+} \leq Ru(bpy)(\mu\text{-}2,3\text{-}dpp)_2^{2+} < Ru(\mu\text{-}2,3\text{-}dpp)_3^{2+}$$

Thus, the synthetic control translates into a high degree of control on the direction of energy flow within these molecules. For the well-known correlation between electrochemical and photophysical properties in these polypyridine complexes,[34,36] the above series is the same as that reported in Section IV.B.

In the case of tetranuclear compounds belonging to the polypyridine family, all four possible energy migration patterns, schematized in Figure 21, have been obtained.[20,58] Pattern (i) is found for L = bpy, BL = 2,3-dpp, and M = Ru^{2+}. In such a complex, the three peripheral units are equivalent and their lowest excited state lies at lower energy than the lowest excited state of the central unit.[17,47g] The reverse

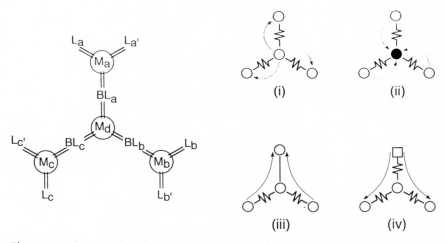

Figure 21. Energy-migration patterns in tetranuclear compounds. Empty and full labels indicate Ru^{2+} and Os^{2+}, respectively. In the peripheral positions, circles and squares indicate M(bpy)$_2$ and M(biq)$_2$ components, respectively. For the bridging ligands: —W— = 2,3-dpp; — = 2,5-dpp.

symmetric pattern (ii) is observed for L = bpy, BL = 2,3-dpp, $M_d = Os^{2+}$, and $M_a = M_b = M_c = Ru^{2+}$.[17] To channel energy towards a single peripheral unit [pattern (iii)], a compound with L = bpy, BL_a = 2,5-dpp, $BL_b = BL_c$ = 2,3-dpp, M = Ru^{2+} was designed and synthesized.[19] Since 2,5-dpp is easier to reduce than 2,3-dpp, the Ru→BL CT excited state of the M_a-based building block is lower in energy than the Ru→BL CT state of the building blocks based on M_b and M_c. On the other hand, the lowest excited state of the central building block is higher in energy than the lowest excited state of all the peripheral ones. Thus, energy transfer from the M_d-, M_b-, and M_c-based building blocks to the M_a-based one is exergonic, but the energy transfer process from the M_b- and M_c-based units to the M_a-based one must either occur directly (coulombic mechanism) or overcome a barrier at M_d (electron-exchange mechanism). The luminescence results show that the only emitting level is that based on M_a and that energy transfer is almost 100% efficient. The migration pattern shown by (iv) was obtained for a compound with $L_a = L_{a'}$ = biq, $L_b = L_{b'} = L_c = L_{c'}$ = bpy, BL = 2,3-dpp, and M = Ru^{2+}.[20] Since the energy level of the M_d-based building block is (slightly) higher than that of the M_a-based one, the migration process must again overcome an energy barrier.

By using different combination of metals and ligands, a variety of energy-migration patterns have also been obtained for hexanuclear[23] and decanuclear (Figure 22)[27] complexes of the same family.

As we have seen in Section III.A, the "complexes-as-metals/complexes-as-ligands" synthetic approach can be extended to obtain even larger dendrimers. Most

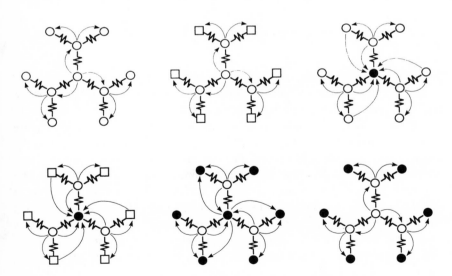

Figure 22. Schematic representation of the energy transfer processes in the decanuclear compounds.[27] For the graphic symbols, see caption to Figure 21.

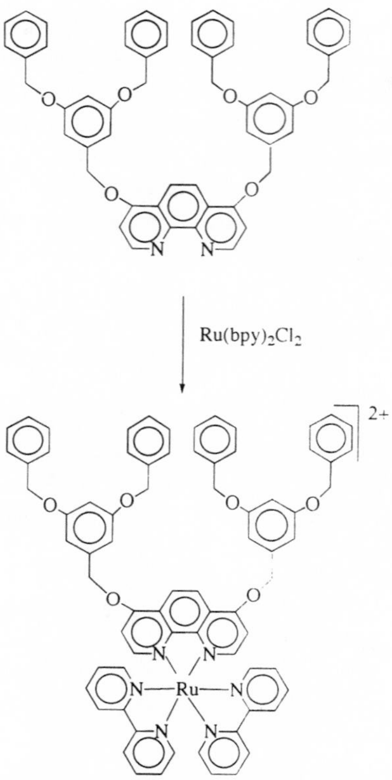

Figure 23. Structure of a compound made by a metal-complex core and appended organic arborols.[72]

of the work has so far been focused on compounds containing Ru(II) and Os(II). However, synthetic control of the energy gradient in larger structures, in order to obtain energy transfer according to predetermined patterns, will probably require the use of more than two different metals. By using cyclometallating ligands, the scope of the "complexes-as-metals/complexes-as-ligands" synthetic strategy can be noticeably extended. A first step along this direction has been the synthesis and characterization of four tetranuclear Ru–Rh, Ru–Ir, Os–Rh, and Os–Ir complexes (**4M-P**, Scheme 1 and Table 1) involving Rh(III)- and Ir(III)-based units which contain the cyclometallating phenylpyridine anion (ppy⁻) as terminal ligand.[21] A further step has led to the synthesis of the hexanuclear compound **6I** (Scheme 1 and Table 1) based on a bridging ligand that carries both N–N and C⁻–N chelating sites.[24]

    An alternative approach to the construction of a light-harvesting antenna system is to synthesize organic "arborols" containing a metal chelating site. With these kind of ligands it is possible to obtain organic chromophores around a metal complex unit.[32a,g,72] An attempt in this direction is schematized in Figure 23.[72] The advantage of these dendrimers compared to the fully organic ones is the presence of a reversible redox center. Organic chromophores absorbing in the visible region will be needed for solar energy conversion.

V. CONCLUSIONS AND PERSPECTIVES

In the last few years a great number of polynuclear transition metal complexes have been synthesized.[39] Such compounds usually have well-defined composition and, in several cases, present a modular structure since they can contain repetitive metal-based components and/or spacers made of repetitive units. Research in the field of polynuclear transition metal complexes can give an important contribution to the development of modular chemistry[46] and to the bottom-up design of nano-structures.[13]

When interaction between the metal-based components is weak, polynuclear transition metal complexes belong to the field of supramolecular chemistry. At the roots of supramolecular chemistry is the concept that supramolecular species have the potential to achieve much more elaborated tasks than simple molecular components; while a molecular component can be involved in simple *acts*, supramolecular species can perform *functions*.[1-5] In other words, supramolecular species have the potentiality to behave as molecular devices. Particularly interesting molecular devices are those which use light to achieve their functions. Molecular devices which perform *light-induced functions* are called *photochemical molecular devices* (PMD).[73] Luminescent and redox-active polynuclear complexes as those described in this chapter can play a role as PMDs operating by photoinduced energy and electron transfer processes.[39]

The interest in highly branched polynuclear metal complexes, and more generally in dendritic species, is related not so much to their size, but rather to the presence of different components. An ordered array of different components can in fact generate valuable properties, such as the presence of cavities having different size, surfaces with specific functions, gradients for photoinduced directional energy and electron transfer, and sites for multielectron transfer catalysis. Studies along these directions are underway in our laboratories.

ACKNOWLEDGMENTS

This work has been supported by the Ministero della Università e della Ricerca Scientifica e Tecnologica and by the Consiglio Nazionale delle Ricerche (Progetto Strategico Tecnologie Chimiche Innovative).

REFERENCES AND NOTES

1. *Molecular Electronics Devices*; Carter, F. L.; Siatkowski, R. E.; Wohltjen, H., Eds.; North-Holland: Amsterdam, 1988; Bard, A. J. *Integrated Chemical Systems*; Wiley: New York, 1995.

2. Drexler, K. E. *Nanosystems. Molecular Machinery, Manufacturing, and Computation*; Wiley: New York, 1992.

3. Balzani, V.; Scandola, F. *Supramolecular Photochemistry*; Horwood: Chichester, 1991.

4. Vögtle, F. *Supramolecular Chemistry*; Wiley: Chichester, 1991.

5. Lehn, J.-M. *Supramolecular Chemistry*; VCH: Weinheim, 1995.

6. For some recent papers, see: (a) Coffin, M. A.; Bryce, M. R.; Batsanov, A. S.; Howard, J. A. K. *J. Chem. Soc., Chem. Commun.* **1993**, 552; (b) Van der Made, A. W.; Van Leeuwen, P. W. N. M.; de Wilde, J.; Brandes, R. A. C. *Adv. Mat.* **1993**, *5*, 466; (c) Mülhaupt, R.; Wörner, C. *Angew. Chem. Int. Ed. Engl.* **1993**, *32*, 1306; (d) de Brabander-van den Berg, E. M. M.; Meijer, E. W. *Angew. Chem. Int. Ed. Engl.* **1993**, *32*, 1308; (e) Xu, Z.; Moore, J. S. *Angew. Chem. Int. Ed. Engl.* **1993**, *32*, 1354; (f) Moulines, F.; Djakovitch, L.; Boese, R.; Gloaguen, B.; Thiel, W.; Fillaut, J.-L.; Delville, M.-H.; Astruc, D. *Angew. Chem. Int. Ed. Engl.* **1993**, *32*, 1075; (g) Newkome, G. R.; Moorefield, C. N.; Keith, J. M.; Baker, G. R.; Escamilla, G. H. *Angew. Chem. Int. Ed. Engl.* **1994**, *33*, 666; (h) Hawker, C. J.; Wooley, K. L.; Fréchet, J. M. J. *J. Chem. Soc., Chem. Commun.* **1994**, 925; (i) Nagasaki, T.; Kimura, O.; Ukon, M.; Arimori, S.; Hamachi, I.; Shinkai, S. *J. Chem. Soc., Perkin Trans. 1* **1994**, 75; (j) Kadei, K.; Moors, R.; Vögtle, F. *Chem. Ber.* **1994**, *127*, 897; (k) Dandlinker, P. J.; Diederich, F.; Gross, M.; Knobler, C. B.; Louati, A.; Sandford, E. M. *Angew. Chem. Int. Ed. Engl.* **1994**, *33*, 1739; (l) Percec, V.; Chu, P.; Ungar, G.; Zhou, J. *J. Am. Chem. Soc.* **1995**, *117*, 11441; (m) Duan, R. G.; Miller, L. L.; Tomalia, D. A. *J. Am. Chem. Soc.* **1995**, *117*, 10783.

7. Buhleier, E.; Wehner, W.; Vögtle, F. *Synthesis* **1978**, 155.

8. Newkome, G. R.; Yao, Z.-Q.; Baker, G. R.; Gupta, V. K. *J. Org. Chem.* **1985**, *50*, 2003.

9. Tomalia, D. A.; Baker, H.; Dewald, J. R.; Hall, M.; Kallos, G.; Martin, S.; Roeck, J.; Ryder, J.; Smith, P. *Polymer J.* **1985**, *17*, 117.

10. (a) Tomalia, D. A.; Naylor, A. M.; Goddard III, W. A. *Angew. Chem. Int. Ed. Engl.* **1990**, *29*, 138; (b) Newkome, G. R.; Moorefield, C. N.; Baker, G. R. *Aldrichimica Acta* **1992**, *25*, 31; (c) Tomalia, D. A.; Durst, H. D. *Top. Curr. Chem.* **1993**, *165*, 193; (d) Fréchet, J. M. J. *Science* **1994**, *263*, 1710; (e) Ardoin, N.; Astruc, D. *Bull. Soc. Chim. Fr.* **1995**, *132*, 875; (f) Issberner, J.; Moors, R.; Vögtle, F. *Angew. Chem. Int. Ed. Engl.* **1994**, *33*, 2413.

11. (a) Mekelburger, H.-B.; Jaworek, W.; Vögtle, F. *Angew. Chem. Int. Ed. Engl.* **1992**, *31*, 1571; (b) Dagnini, R. *Chem. Eng. News* **1993**, *February 1*, 28; (c) Fox, M. A.; Jones Jr., W. E.; Watkins, D. M. *Chem. Eng. News* **1993**, *March 15*, 38; (d) O' Sullivan, D. A. *Chem. Eng. News* **1993**, *August 16*, 20; (e) Hodge, P. *Nature* **1993**, *362*, 18; (f) Tomalia, D. A.; Dvornic, P. R. *Nature* **1994**, *372*, 617; (g) Service, R. F. *Science* **1995**, *267*, 458; (h) Bradley, D. *Science* **1995**, *270*, 1924.

12. Denti, G.; Serroni, S.; Campagna, S.; Juris, A.; Ciano, M.; Balzani, V. In *Perspectives in Coordination Chemistry*; Williams, A. F.; Floriani, C.; Merbach, A. E., Eds.; VCH: Basel, Switzerland, 1992, p. 153.

13. Denti, G.; Serroni, S.; Campagna, S.; Juris, A.; Balzani, V. *Molec. Cryst. Liq. Cryst.* **1993**, *234*, 79.

14. Denti, G.; Campagna, S.; Balzani, V. In *Mesomolecules: from Molecules to Materials*; Mendenhall, D.; Greenberg, A.; Liebman, J., Eds.; Chapman and Hall: New York, 1995, p. 69.

15. Balzani, V.; Denti, G.; Serroni, S.; Campagna, S.; Ricevuto, V.; Juris, A. *Indian Acad. Sciences* **1993**, *105*, 1.

16. Balzani, V.; Campagna, S.; Denti, G.; Juris, A.; Serroni, S.; Venturi, M. *Coord. Chem. Rev.* **1994**, *132*, 1.

17. Campagna, S.; Denti, G.; Sabatino, L.; Serroni, S.; Ciano, M.; Balzani, V. *J. Chem. Soc., Chem. Commun.* **1989**, 1500.

18. Denti, G.; Campagna, S.; Sabatino, L.; Serroni, S.; Ciano, M.; Balzani, V. *Inorg. Chem.* **1990**, *29*, 4750.

19. Denti, G.; Serroni, S.; Campagna, S.; Ricevuto, V.; Balzani, V. *Inorg. Chim. Acta* **1991**, *182*, 127.

20. Denti, G.; Serroni, S.; Campagna, S.; Ricevuto, V.; Balzani, V. *Coord. Chem. Rev.* **1991**, *111*, 227.

21. Serroni, S.; Juris, A.; Campagna, S.; Venturi, M.; Denti, G.; Balzani, V. *J. Am. Chem. Soc.* **1994**, *116*, 9086.

22. Campagna, S.; Denti, G.; Serroni, S.; Ciano, M.; Balzani, V. *Inorg. Chem.* **1991**, *30*, 3728.

23. Denti, G.; Serroni, S.; Campagna, S.; Ricevuto, V.; Juris, A.; Ciano, M.; Balzani, V. *Inorg. Chim. Acta* **1992**, *198–200*, 507.

24. Campagna, S.; Serroni, S.; Juris, A.; Venturi, M.; Balzani, V. *New J. Chem.*, in press.

25. Denti, G.; Campagna, S.; Sabatino, L.; Serroni, S.; Ciano, M.; Balzani, V. *Inorg. Chim. Acta* **1990**, *176*, 175.

26. Serroni, S.; Denti, G.; Campagna, S.; Ciano, M.; Balzani, V. *J. Chem. Soc., Chem. Commun.* **1991**, 944.

27. Denti, G.; Campagna, S.; Serroni, S.; Ciano, M.; Balzani, V. *J. Am. Chem. Soc.* **1992**, *114*, 2944.

28. Campagna, S.; Denti, G.; Serroni, S.; Ciano, M.; Juris, A.; Balzani, V. *Inorg. Chem.* **1992**, *31*, 2982.

29. Serroni, S.; Denti, G.; Campagna, S.; Juris, A.; Ciano, M.; Balzani, V. *Angew. Chem. Int. Ed. Engl.* **1992**, *31*, 1493.

30. Campagna, S.; Denti, G.; Serroni, S.; Juris, A.; Venturi, M.; Ricevuto, V.; Balzani, V. *Chem. Eur. J.* **1995**, *1*, 211.

31. Dendrimers containing metal atoms have also been prepared by other groups.[32] In this article we will only discuss the family of dendrimers prepared in our laboratory.

32. (a) Newkome, G. R.; Cardullo, F.; Constable, E. C.; Moorefield, C. N.; Cargill Thompson, A. M. W. *J. Chem. Soc., Chem. Commun.* **1993**, 925; (b) Newkome, G. R.; Moorefield, C. N. *Macromol. Symp.* **1994**, *77*, 63; (c) Cloutet, E.; Fillaut, J.-L.; Gnanou, Y.; Astruc, D. *J. Chem. Soc., Chem. Commun.* **1994**, 2433; (d) Achar, S.; Puddephatt, R. J. *Angew. Chem. Int. Ed. Engl.* **1994**, *33*, 847; (e) Liao, Y.-H.; Moss, J. R. *Organometallics* **1995**, *14*, 2130; (f) Huck, W. T. S.; van Veggel, F. C. J. M.; Kropman, B. L.; Blank, D. H. A.; Keim, E. G.; Smithers, M. M. A.; Reinhoudt, D. N. *J. Am. Chem. Soc.* **1995**, *117*, 8293; (g) Newkome, G. R.; Güther, R.; Moorefield, C. N.; Cardullo, F.; Echegoyen, L.; Perez-Cordero, E.; Luftmann, H. *Angew. Chem. Int. Ed. Engl.* **1995**, *34*, 2023.

33. Balzani, V.; Bolletta, F.; Gandolfi, M. T.; Maestri, M. *Top. Curr. Chem.* **1978**, *75*, 1.

34. (a) De Armond, M. K.; Carlin, C. M. *Coord. Chem. Rev.* **1981**, *36*, 325; (b) Dodsworth, E. S.; Vlcek, A. A.; Lever, A. B. P. *Inorg. Chem.* **1994**, *33*, 1045.

35. Meyer, T. J. *Pure Appl. Chem.* **1986**, *58*, 1193.

36. Juris, A.; Balzani, V.; Barigelletti, F.; Campagna, S.; Belser, P.; von Zelewsky, A. *Coord. Chem. Rev.* **1988**, *84*, 85.

37. Kalyanasundaram, K. *Photochemistry of Polypyridine and Porphyrin Complexes*; Academic Press: London, 1992.

38. Scandola, F.; Indelli, M. T.; Chiorboli, C.; Bignozzi, C. A. *Top. Curr. Chem.* **1990**, *158*, 73.

39. Balzani, V.; Juris, A.; Venturi, M.; Campagna, S.; Serroni, S. *Chem. Rev.* **1996**, *96*, 759.

40. Balzani, V.; Carassiti, V. *Photochemistry of Coordination Compounds*; Academic Press: London, 1970.

41. Crosby, G. A. *Acc. Chem. Res.* **1975**, *8*, 231.

42. Balzani, V.; Barigelletti, F.; De Cola, L. *Top. Curr. Chem.* **1990**, *158*, 31.

43. Horváth, O.; Stevenson, K. L. *Charge Transfer Photochemistry of Coordination Compounds*; VCH: New York, 1993.

44. Maestri, M.; Balzani, V.; Deuschel-Cornioley, C.; von Zelewsky, A. *Adv. Photochem.* **1992**, *17*, 1.

45. Serroni, S.; Denti, G. *Inorg. Chem.* **1992**, *31*, 4251.

46. *Modular Chemistry*; Michl, J. Ed.; Kluwer: Dordrecht, The Netherlands, 1996, in press.

47. For binuclear and trinuclear complexes based on the 2,3- and 2,5-dpp bridging ligands prepared by our group, see: (a) Campagna, S.; Denti, G.; De Rosa, G.; Sabatino, L.; Ciano, M.; Balzani, V. *Inorg. Chem.* **1989**, *28*, 2565; (b) Campagna, S.; Denti, G.; Sabatino, L.; Serroni, S.; Ciano, M.; Balzani, V. *Gazz. Chim. Ital.* **1989**, *119*, 415; (c) Denti, G.; Serroni, S.; Sabatino, L.; Ciano, M.; Ricevuto, V.; Campagna, S. *Gazz. Chim. Ital.* **1991**, *121*, 37; (d) Juris, A.; Venturi, M.; Pontoni, L.; Resino, I.; Balzani, V.; Serroni, S.; Campagna, S.; Denti, G. *Can. J. Chem.* **1995**, *73*, 1875. For binuclear, trinuclear, and tetranuclear compounds based on the 2,3-dpp bridging ligand prepared by other groups, see: (e) Fuchs, Y.; Lofters, S.; Dieter, T.; Shi, W.; Morgan, R.; Strekas, T. C.; Gafney, H. D.; Baker, A. D. *J. Am. Chem. Soc.* **1987**, *109*, 2691; (f) Ernst, S.; Kasack, V.; Kaim, W. *Inorg. Chem.* **1988**, *27*, 1146; (g) Murphy, W. R.; Brewer, K. J.; Gettliffe, G.; Petersen, J. D. *Inorg. Chem.* **1989**, *28*, 81; (h) Berger, R. M. *Inorg. Chem.* **1990**, *29*, 1920; (i) Cooper, J. B.; MacQueen, D. B.; Petersen, J. D.; Wertz, D. W. *Inorg. Chem.* **1990**, *29*, 3701; (j) Richter, M. M.; Brewer, K. J. *Inorg. Chem.* **1992**, *31*, 1594; (k) Kalyanasundaram, K.; Graetzel, M.; Nazeeruddin, Md. K. *J. Phys. Chem.* **1992**, *96*, 5865.

48. Buchanan, B. E.; Wang, R.; Vos, J. G.; Hage, R.; Haasnoot, J. G.; Reedijk, J. *Inorg. Chem.* **1990**, *29*, 3263.

49. (a) Didier, P.; Jacquet, L.; Kirsch-De Mesmaeker, A.; Hueber, R.; van Dorsselaer, A. *Inorg. Chem.* **1992**, *31*, 4803; (b) Denti, G.; Serroni, S.; Sindona, G.; Uccella, N. *J. Am. Soc. Mass Spectrom.* **1993**, *4*, 306.

50. Arakawa, R.; Matsuo, T.; Ohno, T.; Haga, M. *Inorg. Chem.* **1995**, *34*, 2464.

51. Moucheron, C.; Dietrich-Buchecker, C. O.; Sauvage, J.-P.; Van Dorsselaer, A. *J. Chem. Soc., Dalton Trans.* **1994**, 885. Russel, K. C.; Leize, E.; Van Dorsselaer, A.; Lehn, J.-M. *Angew. Chem. Int. Ed. Engl.* **1995**, *34*, 209.

52. (a) Orellana, G.; Kirsch-De Mesmaeker, A.; Turro, N. J. *Inorg. Chem.* **1990**, *29*, 882; (b) Marzin, C.; Budde, F.; Steel, P. J.; Lerner, D. *New J. Chem.* **1987**, *11*, 33; (c) Brevard, C.; Granger, P. *Inorg. Chem.* **1983**, *22*, 532; (d) Steel, P. J.; Lahousse, F.; Lerner, D.; Marzin, C. *Inorg. Chem.* **1983**, *22*, 1488.

53. Predieri, G.; Vignali, C.; Denti, G.; Serroni, S. *Inorg. Chim. Acta* **1993**, *205*, 145.

54. Exception being made for the tetranuclear complexes described in Ref. 21 that are unstable in acetonitrile and other coordinating solvents.

55. Campagna, S.; Giannetto, A.; Serroni, S.; Denti, G.; Trusso, S.; Mallamace, F.; Micali, N. *J. Am. Chem. Soc.* **1995**, *117*, 1754.

56. See, for example, Hage, R.; Dijkhnis, A. H. J.; Haasnoot, J. G.; Prins, R.; Reedijk, J.; Buckanan, B. E.; Vos, J. G. *Inorg. Chem.* **1988**, *27*, 2185.

57. Denti, G.; Campagna, S.; Sabatino, L.; Serroni, S.; Ciano, M.; Balzani, V. In *Photochemical Conversion and Storage of Solar Energy*; Pelizzetti, E.; Schiavello, M., Eds.; Kluwer: Dordrecht, 1991, p. 27.

58. Balzani, V.; Campagna, S.; Denti, G.; Serroni, S. In *Photoprocesses in Transition Metal Complexes, Biosystems, and other Molecules*; Kochanski, E., Ed.; Kluwer: Dordrecht, 1992, p. 233.

59. Juris, A.; Balzani, V.; Campagna, S.; Denti, G.; Serroni, S.; Frei, G.; Güdel, H. U. *Inorg. Chem.* **1994**, *33*, 1491.

60. Campagna, S.; Denti, G.; Serroni, S.; Juris, A.; Venturi, M.; Balzani, V. In *Self-production of Supramolecular Species*; Fleischaker, G. R.; Colonna, S.; Luisi, P. L., Eds.; Kluwer: Dordrecht, 1994, p. 261.

61. Balzani, V.; Campagna, S.; Denti, G.; Juris, A.; Serroni, S.; Venturi, M. *Solar En. Mater. Solar Cells* **1995**, *38*, 159.

62. Roffia, S.; Marcaccio, M.; Paradisi, C.; Paolucci, F.; Balzani, V.; Denti, G.; Serroni, S.; Campagna, S. *Inorg. Chem.* **1993**, *32*, 3003.

63. Flanagan, J. B.; Margel, S.; Bard, A. J.; Anson, F. C. *J. Am. Chem. Soc.* **1978**, *100*, 4248.

64. Campagna, S.; Ricevuto, V.; Denti, G.; Serroni, S.; Juris, A.; Ciano, M.; Balzani, V. In *Advanced Syntheses and Methodologies in Inorganic Chemistry*; Daolio, S.; Fabrizio, M.; Guerriero, P.; Tondello, E.; Vigato, P. A., Eds.; University Press: Padova, 1992, p. 128.

65. Barigelletti, F.; De Cola, L.; Balzani, V.; Hage, R.; Haasnoot, J. G.; Reedijk, J.; Vos, J. G. *Inorg. Chem.* **1989**, *28*, 4344.

66. Astruc, D. *Electron Transfer and Radical Processes in Transition-Metal Complexes*; VCH: New York, 1995.

67. Charge transfer transitions between remote centers can also be expected, but they give rise to much weaker absorptions; see, e.g., Bignozzi, C. A.; Paradisi, C.; Roffia, S.; Scandola, F. *Inorg. Chem.* **1988**, *27*, 408.

68. (a) Bradley, P. C.; Kress, N.; Hornberger, B. A.; Dallinger, R. F.; Woodruff, W. H. *J. Am. Chem. Soc.* **1989**, *103*, 7441; (b) Cooley, L. F.; Bergquist, P.; Kelley, D. F. *J. Am. Chem. Soc.* **1990**, *112*, 2612.

69. (a) Balzani, V.; Credi, A.; Scandola, F. In *Transition Metals in Supramolecular Chemistry*; Fabbrizzi, L.; Poggi, A., Eds.; Kluwer: Dordrecht, The Netherlands, 1994, p. 1; (b) Balzani, V.; Scandola, F. In *Comprehensive Supramolecular Chemistry*; Reinhoudt, D. N., Ed.; Pergamon, 1996, Vol. 10, p. 1.

70. (a) Jones, W. E.; Baxter, S. M.; Strouse, G. F.; Meyer, T. J. *J. Am. Chem. Soc.* **1993**, *115*, 7363; (b) Jones, W. E.; Baxter, S. M.; Mecklenburg, S. L.; Erickson, B. W.; Peek, B. M.; Meyer, T. J. In *Supramolecular Chemistry*; Balzani, V.; De Cola, L., Eds.; Kluwer: Dordrecht, The Netherlands, 1992, p. 249.

71. (a) Wagner, R. W.; Lindsey, J. S. *J. Am. Chem. Soc.* **1994**, *116*, 9759; (b) Seth, J.; Palaniappan, V.; Johnson, T. E.; Prathapan, S.; Lindsey, J. S.; Bocian, D. F. *J. Am. Chem. Soc.* **1994**, *115*, 10578; (c) Anderson, S.; Anderson, D. H.; Bashall, A.; McPartlin, M.; Sanders, J. K. M. *Angew. Chem. Int. Ed. Engl.* **1995**, *34*, 1096.

72. Serroni, S.; Campagna, S.; Juris, A.; Venturi, M.; Balzani, V.; Denti, G. *Gazz. Chim. Ital.* **1994**, *124*, 423.

73. (a) Balzani, V.; Moggi, L.; Scandola, F. In *Supramolecular Photochemistry*; Balzani, V., Ed.; Reidel: Dordrecht, The Netherlands, 1987, p. 1; (b) Balzani, V.; Scandola, F. *Supramolecular Photochemistry*; Horwood: Chichester, 1991, Chapter 12.

REDOX-ACTIVE DENDRIMERS, RELATED BUILDING BLOCKS, AND OLIGOMERS

Martin R. Bryce and Wayne Devonport

Advances in Dendritic Macromolecules
Volume 3, pages 115–149
Copyright © 1996 by JAI Press Inc.
All rights of reproduction in any form reserved.
ISBN: 0-7623-0069-8

ABSTRACT

This chapter reviews progress on redox-active dendrimers, and includes literature published up to the end of 1995. Understanding the electron transfer processes in large macromolecules can be very challenging, especially where multiple redox sites are present, so studies on model branched oligomers and monomeric building blocks are also discussed. Selected synthetic schemes are outlined. The division of subject matter is based on the structure of the redox-active units, which are: ferrocene and related transition metal sandwiches; pyridine-based transition metal complexes; tetrathiafulvalene derivatives; porphyrin and phthalocyanine derivatives; fullerenes; poly(amidoamine) derivatives; poly(arylene)methanes; and cobalt(salen) analogs. The electrochemical and chemical redox properties of these materials will be emphasized. Their relevance in such fields as redox catalysts, organic conductors, modified electrodes, and models for electron transfer proteins are outlined, with reference to specific materials.

I. INTRODUCTION

The initial report by Vögtle et al. in 1978 on the preparation and characterization of polyamines with branched topologies around a trigonal nitrogen center via iterative (cascade) synthetic methodology[1] sparked an explosion of interest in highly branched, tree-like molecules, variously called cascade molecules, arborols, and dendrimers. Authoritative general reviews on this subject have been published recently by Newkome et al.,[2] Fréchet,[3] and Tomalia et al.[4] Unlike many other synthetic macromolecules, e.g. linear, step-growth polymers, these materials possess a high degree of order within the structural architecture. For example, control over molecular weight, topology, interior cavity size, and surface functionality can be readily achieved.

These macromolecules are very attractive to scientists from a wide range of disciplines, not only because of their aesthetic appeal, the fundamental synthetic challenges they pose, and the unusual physical and chemical properties which they display, but also because of their potential applications[5] in such diverse fields as gene therapy, hosts for the transport of biologically important guests, engineering plastics, redox catalysis, liquid crystals, micellar substitutes for chromatographic separations,[6] and molecular electronic devices. Interior cavities might even serve as nanoscale reaction chambers for small guest molecules.[4b] Within the last few years, emphasis has shifted away from the synthesis of dendrimers of ever-increasing size[7] towards the incorporation of functional components within the branched skeletons.[8]

II. FUNCTIONAL DENDRIMERS

A range of more elaborate functional groups have recently been appended to, or embedded within, the dendrimer framework. These derivatives are designed to possess special properties. They include: (1) crown ethers that can complex metal ions within the core and around the periphery;[9] (2) chiral groups at the nucleus and within the branches for enantiomeric host-guest recognition;[10] (3) liquid crystalline derivatives;[11] (4) polyammonium cation dendrimers for catalytic applications;[12] (5) polynuclear metal complexes which are luminescent;[13] (6) dendrimers with internal cavities and receptor sites which can act as hosts guest molecules,[14] including systems with closed surfaces in which the guests were encapsulated during the synthesis;[14b] and (7) redox-active systems. This last topic is the focus of this review.

III. REDOX-ACTIVE SYSTEMS

A variety of redox-active organic and organometallic groups have been incorporated into dendritic and hyperbranched systems with several aims in mind. These include: (1) new electron transfer catalysts; (2) studies on the dynamics of electron transport at surfaces and within restricted reaction spaces;[15] (3) new materials for energy conversion; (4) organic semiconductors; (5) organic magnets; and (6) mimics of biological redox processes.

Some dendrimer systems contain a single redox-active unit at the core for which the key issue is generally to observe how the redox behavior of this central "encapsulated" group is modulated by the shielding effect of the rest of the dendrimer structure. This should provide a more compact insulating layer than would an analogous linear polymer. The majority of studies, however, concern multiple redox units emplaced within the branches and/or at peripheral sites; interest is focussed upon these dendrimers acting as multielectron sources or reservoirs where the redox groups may act independently in multi-electron processes (n identical electroactive centers which undergo electron transfer in a single n-electron wave) or they may interact intra- or intermolecularly (overlapping or closely-spaced redox waves at different potentials). Cooperative phenomena (electron delocalization, magnetism, conductivity, etc.) are possible with these systems if the radical species are sufficiently stable.

The subject matter covered below is divided into sections according to the structure of the redox unit(s). This review is restricted primarily to materials for which well-defined redox behavior has been reported, usually involving cyclic voltammetric studies and other electrochemical techniques in solution. Unraveling the electron transfer processes in larger macromolecules which contain multiple redox sites can be very challenging, thus for some systems model branched oligomers have been studied in detail, and this work will be discussed. Selected synthetic schemes are included to acquaint the reader with the building blocks which are available for the construction of new derivatives, and with the synthetic steps involved.

A. Ferrocene and Related Transition Metal Sandwiches

Ferrocenyl-based polymers[16] are established as useful materials for the modification of electrodes,[17] as electrochemical biosensors,[18] and as nonlinear optical systems.[19] The redox behavior of ferrocene can be tuned by substituent effects and novel properties can result; for example, permethylation of the cyclopentadienyl rings lowers the oxidation potential, and the charge transfer salt of decamethylferrocene with tetracyanoethylene, $[FeCp_2^*]^{+\bullet} [TCNE]^{-\bullet}$, is a ferromagnet below $T_c = 4.8$ K,[20] and electrode surfaces modified with a pentamethylferrocene derivative have been used as sensors for cytochrome c.[21] These diverse properties have provided an added impetus to studies on ferrocene dendrimers.

Cuadrado et al. reported that hydrosilylation of 1,3,5,7-tetramethylcyclotetrasiloxane (1) with four equivalents of vinylferrocene (2) in the presence of catalytic bis(divinyltetramethyldisiloxane)platinum(0) yielded tetraferrocenyl compound 3 in 92% yield (Scheme 1). The cyclic voltammogram of 3 exhibited a single reversible oxidation wave and coulometry established that this wave corresponded to the removal of four electrons per molecule, suggesting that the four ferrocenyl units act as independent, non-interacting redox centers.[22]

Different methodology (viz. treatment of 4 with ferrocenyllithium 5) produced organosilicon dendrimer 6 with a tetrahedral silicon atom at the core and eight peripheral ferrocenyl groups (Scheme 2). The eight ferrocenyl groups of 6 were oxidized independently in a single, reversible, eight-electron wave at $E^{1/2} = + 0.44$ V (vs. SCE) based on cyclic voltammetry, differential pulse voltammetry, and controlled-potential electrolysis experiments.[23] The polycation species oxidatively precipitated from tetrahydrofuran solution yielding adsorbed films on a platinum electrode surface, suggesting that electrodes modified with ferrocene dendrimers will be accessible.

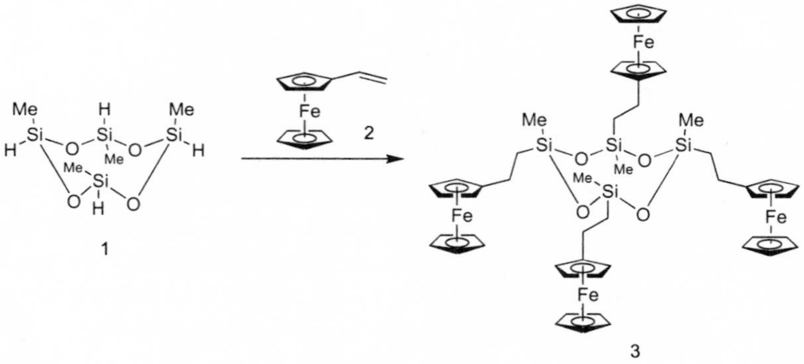

Scheme 1.

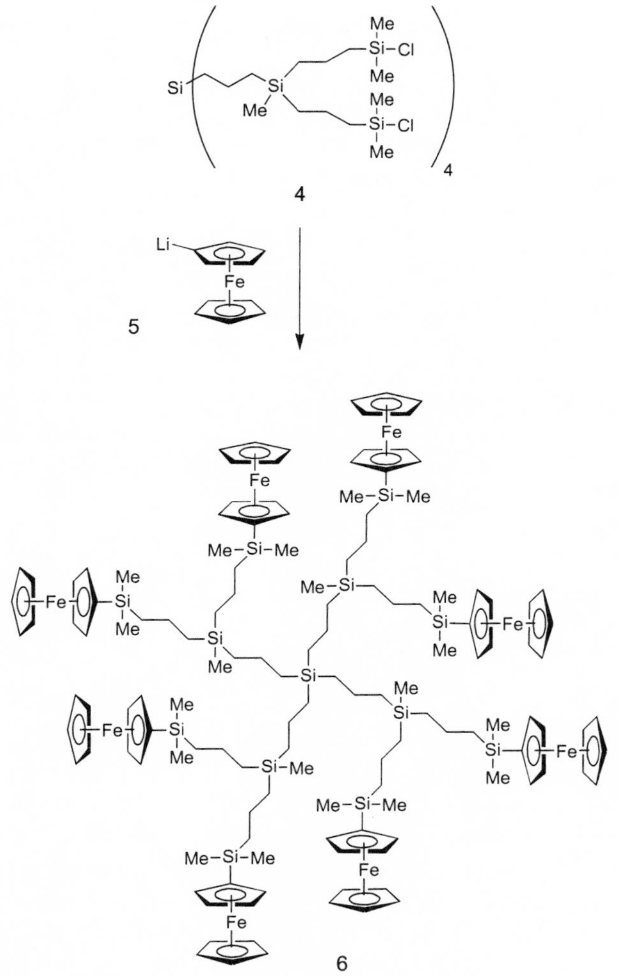

Scheme 2.

Astruc has developed the concept of transition metal sandwiches acting as electron reservoir complexes.[24] The characteristic of an electron reservoir is that the reduced form is easily generated and does not decompose; to increase stability, the radical center can be sterically protected in the heart of a bulky molecular framework. The [FeCp(arene)]$^+$ series of complexes, e.g. **7**, are prime examples, for which variation of the arene structure modulates the redox potential.

Triple-branching functionalization of complex **7** yielded nona-allyl derivative **8** which was demetallated by visible light photolysis in the presence of triphenyl-

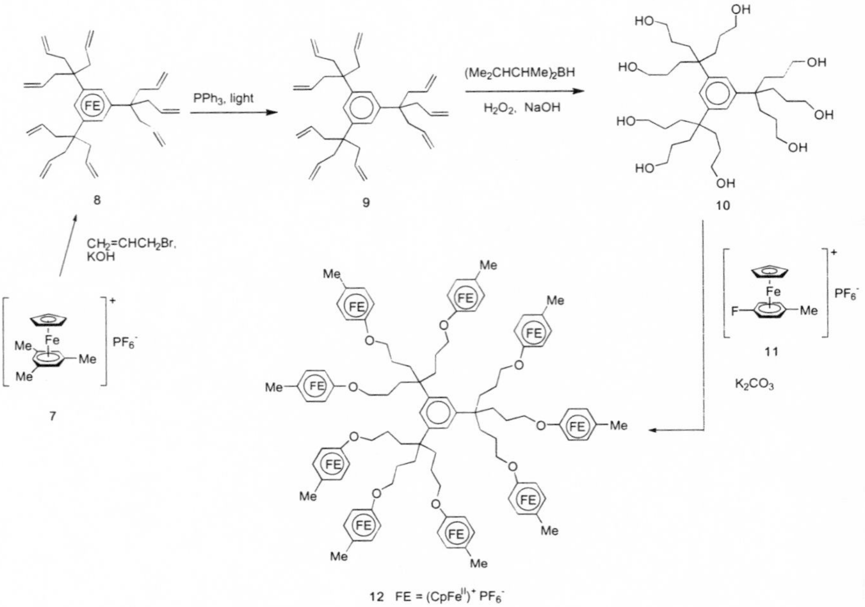

Scheme 3.

phosphine to yield arene **9** (characterized by an X-ray crystal structure). Hydration of **9** (hydroboration, then alkaline oxidation) yielded the hygroscopic nonaol **10**, perarylmetalation of which, using reagent **11** under phase-transfer conditions, afforded the nonairon sandwich complex **12** bearing a +9 charge (Scheme 3).[25] The cyclic voltammogram (CV) of **12** shows a reversible reduction wave for the Fe^{2+} (18 electrons)/Fe^+ (19 electrons) couple at -1.37 V (vs. SCE). From the intensity of the redox wave in the CV the number of electrons transferred per molecule was calculated to be 8 ± 1; thus compound **12** acts as a multielectron reservoir. Heterogeneous electron transfer from the cathode to the closest redox centers in such a system should be faster than to the more remote sites in the molecule. However, the observed kinetics indicated that the rate constants for electron transfer were so similar that the redox sites in **12** were indistinguishable on the electro-chemical timescale.[25]

An extension of this methodology, by hexafunctionalization of the arene group in **13** using ferrocenylbutyl iodide **14** under basic conditions, gave the hexafluoro-phosphate salt of the heptanuclear Fe(II) complex cation **15** in 65% yield.[26] Photolysis of **15**, under similar conditions which converted **8** into **9**, led to isolation of the neutral ferrocenyl complex **16** by selective decomplexation of the central

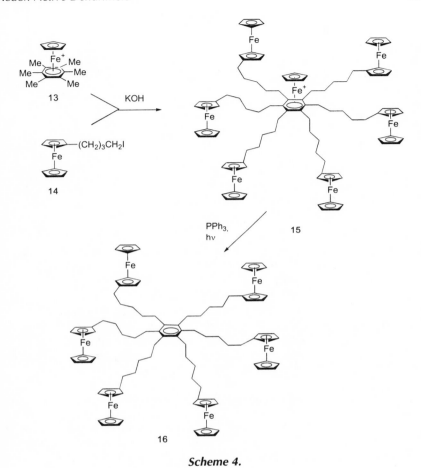

Scheme 4.

cationic FeCp$^+$ unit (Scheme 4). Oxidation of **15** with sulfuric acid, followed by ion exchange, yielded the hexafluorophosphate salt of the corresponding heptanuclear hexaferrocinium derivative as an isolable solid.

The solution electrochemical properties of **15** and **16** were studied.[26] The one-electron wave for the core unit of **15** at negative potential provides a very convenient internal standard, which assists calculations of the number of electrons involved in the redox processes at the periphery. The voltammogram of **15** shows a relative intensity ratio of 6 ± 0.3 between the anodic and cathodic waves which are fully chemically reversible (I_a/I_c = ca. 1.0) (Figure 1). The ΔE_p value of the ferrocene signal (55 mV) indicates the six ferrocene groups are electronically equivalent, and their oxidation potentials are essentially the same in compounds **15** and **16** (E^0 =

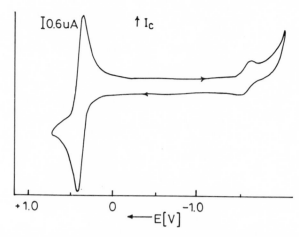

Figure 1. Cyclic voltammogram of compound **15** (solvent DMF, electrolyte n-Bu$_4$N$^+$ PF$_6^-$, Pt electrode, versus SCE) (redrawn from Ref. 26).

+0.44 and +0.45 V, respectively, vs. SCE) indicating that the oxidation process is not influenced by the central cationic unit of **15**. The [FeCp(arene)]$^+$ system is known to be an active catalyst in the multielectron electroreduction of NO$_3^-$ to NH$_3$,[27] thus polyiron sandwiches hold considerable potential as multielectron redox catalysts.[28]

B. Pyridine-Derived Transition Metal Complexes

Polypyridine transition metal complexes[29] possess many attractive features for incorporation into dendrimers: (1) coordination of the ligands to a variety of metals gives stable complexes; (2) they exhibit intense absorption bands in the UV and visible spectra; (3) they display reasonably strong and long-lived luminescence; and (4) they possess a rich redox behavior, comprising reversible one-electron oxidation at the metal center and multielectron reductions at each ligand center.[30] Based on the electrochemical and luminescent behavior of their ruthenium complexes, it was deduced that **17** is a better donor ligand than **18** (the latter is reduced at less negative potentials).[30c] These materials are versatile and new synthetic methodology has enabled specific metals and/or ligands to be placed at predetermined sites in a supramolecular array.

Pyridine-based ligands which have been used for dendrimers are 2,2-bipyridine (bpy) **17**, 2,3-bis(2-pyridyl)pyrazine (2,3-dpp) **18** and its monomethylated salt **19**, and 2,2':6',2''-terpyridine **20**. Their transition metal complexes possessing dendritic structures were first reported in the collaborative work of Denti, Campagne, and Balzani whose divergent synthetic strategy has led to systems containing 22 ruthenium centers.[13,31] The core unit is [Ru(2,3-dpp)$_3$]$^{2+}$ **21** which contains three

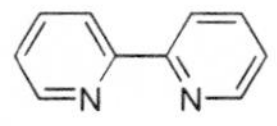

17

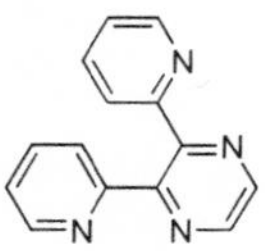

18

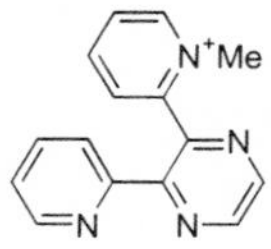

19

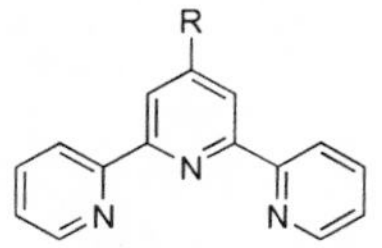

20

R = Functionalized Chain

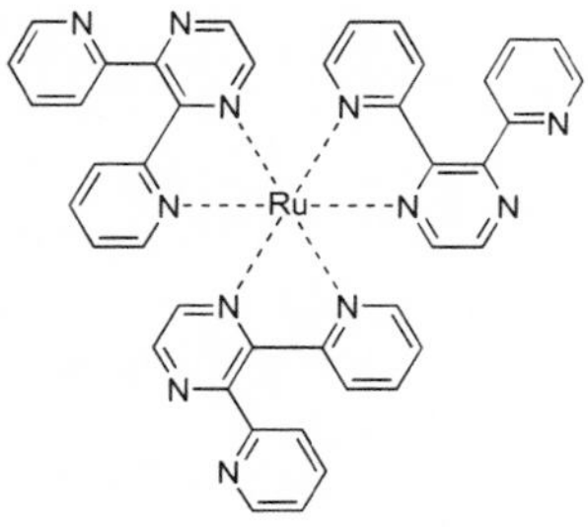

21

vacant sites for coordination. The ruthenium complex of **19** [Ru(2,3-Medpp)$_2$Cl$_2$]$^{2+}$ contains chloride ligands which can be readily replaced, and two bridging ligands in which the chelating sites not used for metal coordination are masked by methylation; demethylation can be achieved by refluxing in the presence of diaz-abicyclo[2.2.2.]octane. The reagent [Ru(2,2′-bipyridyl)$_2$Cl$_2$] provided the periph-eral [Ru(2,2′-bipyridyl)$_2$] groups, which prevented the system from enlarging further. The deca-ruthenium system, thereby obtained, is shown in structure **22**, and an expanded system with 22 ruthenium atoms coordinated to 21 bridging ligands and 24 terminal ligands is represented schematically in structure **23**.

Compounds **22** and **23** are soluble in common organic solvents and stable in light. They display an interesting range of properties including: strong absorption in the UV/visible spectral region, moderately strong red luminescence, and complicated metal-based oxidation and ligand-based reduction processes involving intra-molecular interactions between the different components.[31] The CV and differential pulse voltammetry (DPV) of compound **22** in acetonitrile reveals three waves: one oxidation wave at $E_{OX}^{1/2}$ = +1.53 V, and two broad reduction waves at −0.73 and −1.22 V (vs. SCE), corresponding to six-, six-, and three-electron processes, respectively (Figure 2). The six-electron oxidation is attributed to the oxidation, at nearly the same potential, of the six equivalent peripheral Ru ions. The formation of this positively charged "shell" has the effect of displacing the oxidation potentials of the inner Ru atoms to more positive values, which were outside the range of the solvent window for the experiments. The first six-electron reduction wave is attributed to the one-electron reduction of the six outer bridging ligands, which interact only slightly with one another. The second three-electron reduction wave is assigned to the one-electron reduction of the three inner bridging ligands.

23

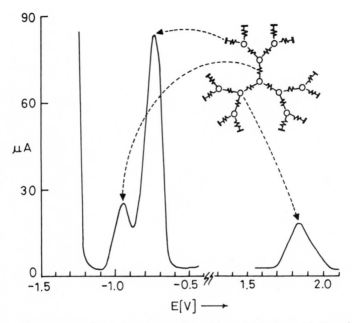

Figure 2. Differential pulse voltammetry of dendrimer **22** (solvent MeCN, electrolyte Et$_4$N$^+$ PF$_6^-$, Pt electrode, versus SCE) and a diagram representing structure **22** (redrawn from Ref. 13).

Compound **23** shows the expected 12-electron oxidation wave at $E_{OX}^{1/2} = +1.52$ V, arising from the 12 equivalent peripheral metals which are considered to be weakly interacting, but the reductive waves were overlapping and could not be assigned.

Luminescence in these materials originates from the peripheral units where the lowest energy excited state is localized. Efficient electronic energy transfer occurs from the inner (higher energy) units to the peripheral (lower energy) ones. As the dendrimers increased in size there was found to be an increase in their ability to absorb sunlight, while the emission lifetimes and quantum yields hardly changed. This suggests that better antennae for light harvesting can be obtained with larger dendrimers.[13]

Dendrimer **23** has been viewed as a potential "organic zeolite" because it could serve as a reaction chamber for small guest molecules.[32] The three-dimensional branched structures of **22** and **23**, as revealed by computer models, are shown in Figures 3a and 3b, respectively.[31]

Terpyridine ligand **20**, with appropriate functionalization at the 4′ position (introduced by nucleophilic displacement of a 4′-chloro substituent) is the key building block used by Newkome and by Constable. The assembly of the beautiful dodecaruthenium macromolecule **24** carrying a 24$^+$ charge was an early highlight

24   R = CH₂Ph

within this class of macromolecules, although its electrochemistry was not reported.[33]

More recently, similar terpyridyl units form the focal point of a series of X<Ru>Y
bis-dendrimers **25** (where X and Y refer to different dendritic components) studied
by Newkome, Echegoyen, and co-workers.[34] Compounds **25a–c** were prepared by
connecting a metallo–terpyridine donor moiety (as the $RuCl_3$ complex) to the
complementary terpyridine receptor moiety (with the metal absent). Evidence for
complexation was provided by NMR and UV spectroscopy, elemental analysis and
solution electrochemistry. System **25a** displayed the expected cathodic and anodic
electrochemistry of the $Ru(terpyridyl)_2^{2+}$ center. As steric hindrance around this
group increased on moving to analog **25b** hardly any change was observed in the
redox behavior of the terpyridyl ligand, whereas the Ru^{2+}/Ru^{3+} couple exhibited a
slightly larger ΔE_{pp} value (the different in peak potentials) which is indicative of a
slower electron transfer process. For compound **25c**, however, all the redox waves

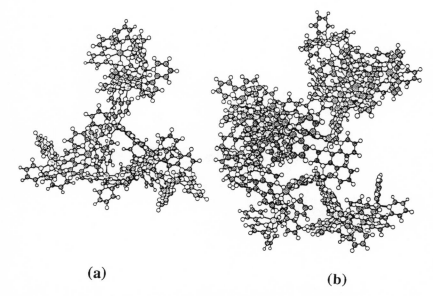

(a)　　　　　　　　　　　　　　　**(b)**

Figure 3. Structures of (**a**) compound **22** and (**b**) compound **23** as revealed by computer models (reproduced from Ref. 31, with the permission of VCH Publishers).

were electrochemically irreversible, and there was also evidence for the onset of chemical irreversibility (small currents were observed upon scan reversal for both cathodic and anodic scans). It was noted that the $E^{1/2}$ values for the redox couples did not change significantly within the series **25a,b,c**, which contrasts with previous work on successive generations of Zn–porphyrin dendrimers (see Section D).

25 (a) R = OC(Me)$_3$
25 (b) R = NHC[CH$_2$CH$_2$CO$_2$C(Me)$_3$]$_3$
25 (c) R = NHC[CH$_2$CH$_2$CONHC(CH$_2$CH$_2$CO$_2$C(Me)$_3$)$_3$]$_3$

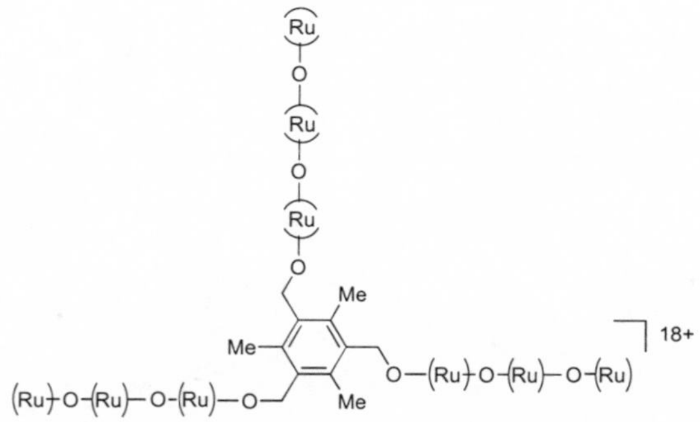

26

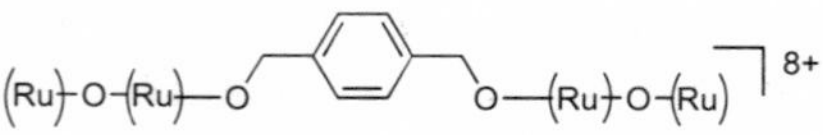

27

System **26**, comprising Ru(terpyridyl)$_2$ units built around an aromatic core by a convergent strategy, showed a single broad oxidation wave in the cyclic voltammogram, whereas the model tetranuclear species **27** exhibited two separate Ru^{2+}/Ru^{3+} redox couples separated by 110 mV, arising from the two different ruthenium environments.[35] Electrochemical studies were hampered by adsorption of the dendrimers onto glassy carbon or platinum electrode surfaces. The complexes exhibited a lowest energy metal-to-ligand charge transfer transition with λ_{max} between 479 and 493 nm, which shifted to the lower energy end of this range with increasing nuclearity. The narrowness of this range is indicative of adjacent ruthenium centers having little influence upon each other due to their separation by the ether linkage.[35b]

Large dendrimers based on the [PtMe$_2$(2,2'-bipyridyl)] unit constructed around a 1,2,4,5-tetrafunctionalized benzene core and containing up to 28 platinum atoms have been reported by Puddephat,[36] along with a related 14 platinum atom system **30** which was obtained by reaction of wedge **28** with the disubstituted 2,2'-bipyridyl derivative **29** (Scheme 5).[37] Progression could not be continued beyond **30** because of steric congestion.

C. Tetrathiafulvalene Derivatives

The incorporation of tetrathiafulvalene (TTF) into dendrimers presents a fascinating prospect for the following reasons: (1) oxidation of TTF to the cation radical and dication species occurs sequentially and reversibly at very accessible potentials (for unsubstituted TTF, $E_1^{1/2} = +0.34$ and $E_2^{1/2} = +0.78$ V, vs. Ag/AgCl); (2) the oxidation potentials can be finely tuned by the attachment of appropriate substi-

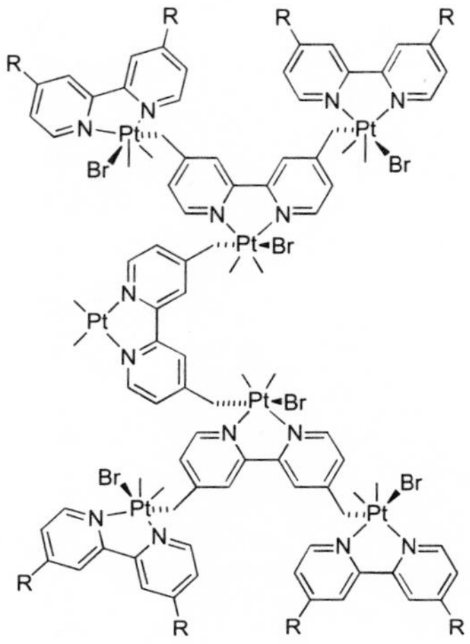

28

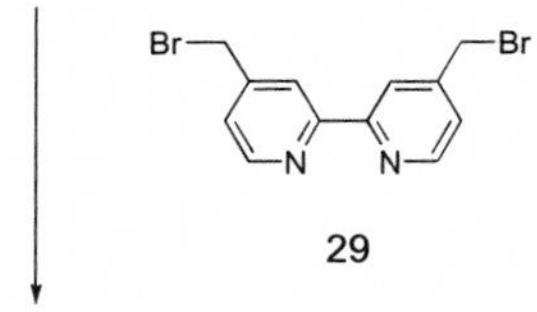

29

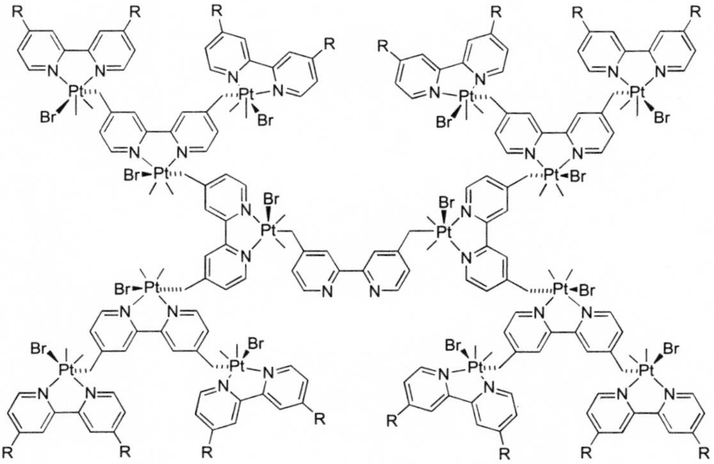

30 R = Bu

Scheme 5.

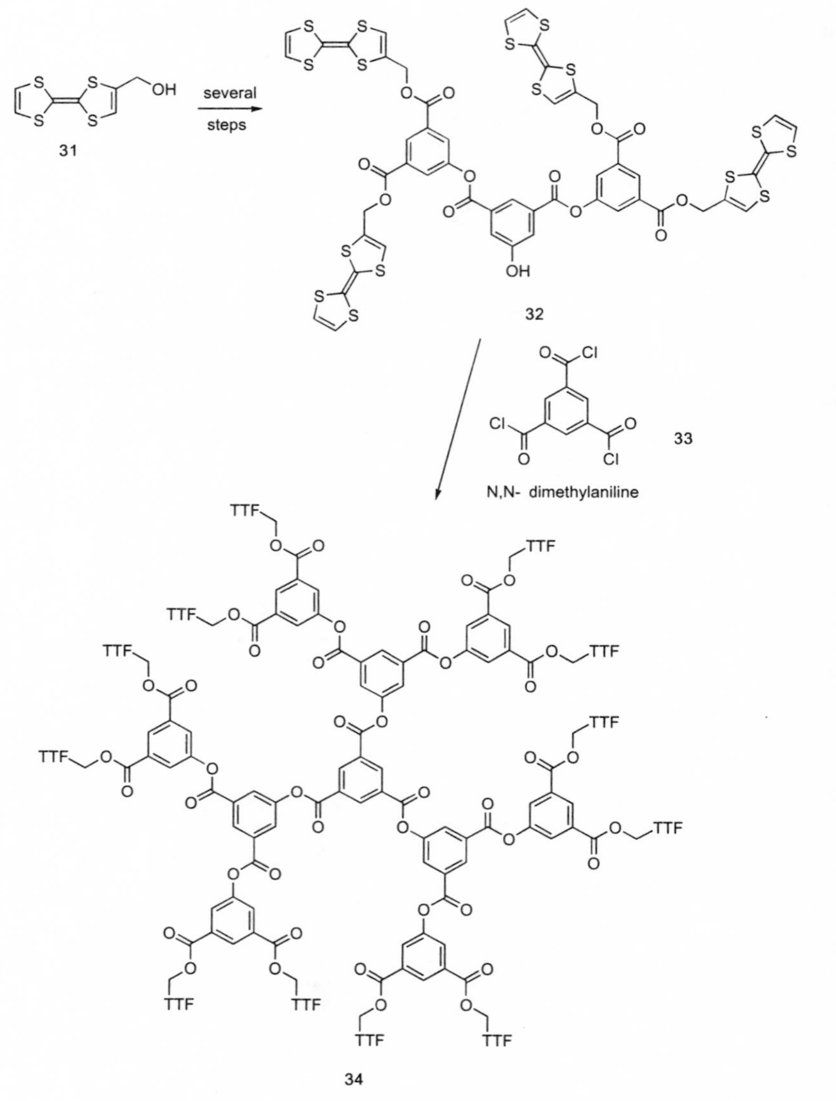

Scheme 6.

tuents; (3) the TTF cation radical is thermodynamically very stable; and (4) oxidized TTF units are key components of molecular conductors because of their propensity to form highly ordered stacks along which there is high electron mobility.[38] The stacks are stabilized by intermolecular π–π interactions and non-bonded sulfur–sulfur interactions. A wide range of TTF dimers and trimers are known,[39] and some main-chain and side-chain TTF polymers have been reported recently by Müllen et al.[40]

Efficient monofunctionalization of TTF via TTF-Li requires precisely controlled reaction conditions to avoid multisubstitution and the formation of inseparable product mixtures.[41] However, the recent availability of many TTF derivatives in multigram quantities[42] presents new opportunities for the use of TTF as a building block in supramolecular chemistry.[43]

TTF dendrimers have been synthesized in our laboratory by a convergent strategy based on a repetitive coupling/deprotection sequence using 4-(hydroxymethyl)-TTF **31**[41] as the starting monomer. We prepared the tetrakis-TTF dendron wedge **32**, which reacted with 1,3,5-benzenetricarbonylchloride (**33**) to furnish dendrimer **34** decorated with 12 TTF units, in 30% overall yield from **31** (Scheme 6).[44] 4,4'-Biphenylether diacidchloride reacted analogously with **32** to yield octakis-TTF system **35** (Scheme 7) possessing a more open structure than analog **34**.[45] While

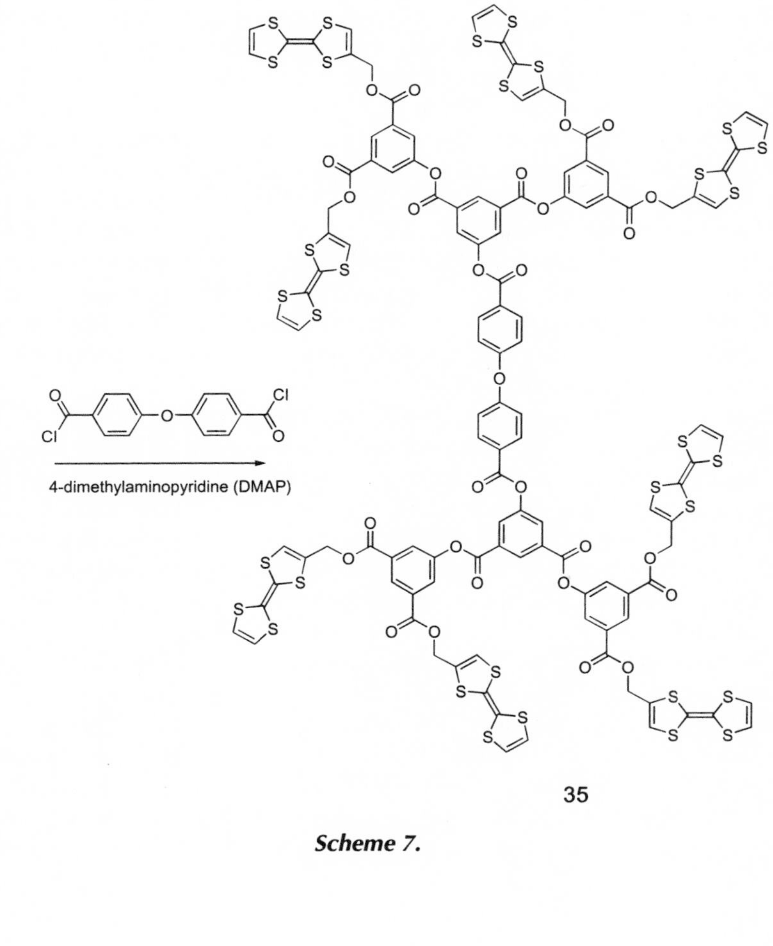

Scheme 7.

compound **34** is stable only when stored at below 0 °C, compound **35** is stable at room temperature for at least one year.

The solution electrochemical redox behavior of systems **34**, **35**, and model multi-TTFs has been studied.[44–46] For compound **34** two redox couples typical of the TTF system were observed at $E_1^{1/2}$ and $E_2^{1/2} = 0.43$ and 0.81 V, respectively (vs. SCE), which are very similar to the values for model monomeric esters of **31**. For lower generation derivatives the redox waves were reversible (the criteria applied were: the separation between the cathodic and anodic peaks ΔE_p, a ratio of unity for the intensities of the cathodic and anodic currents I_c/I_a, and the consistency of the peak potential at various scan rates), whereas for higher-generation systems, e.g. **34**, the waves were quasi-reversible, with slight broadening, which became more noticeable with repeated scans resulting in the deposition of material on the electrode surface. Classical cyclic voltammetry (CV) chronoamperometry and cyclic voltammetry with ultramicroelectrodes (UME CV) (using the reversible, one-electron reduction of 2,3-dichloronaphthoquinone present in a known concentration as an internal reference) showed that each couple represented a multielectron transfer with essentially no interaction between the charged TTF units. The CV data (Figure 4a) implied that each couple involved ca. 5–6 electrons, i.e. only half the expected value of 12 electrons for the 12 TTF groups. We ascribe these discrepencies primarily to diffusion phenomena; in particular, different diffusion rates of the internal standard and the dendrimer could account for the apparent redox activity of only some of the TTF units. Scanning to higher potentials showed no further oxidation waves, which would seem to preclude the presence of any TTF groups buried within the dendrimer structure. The UME CV data (Figure 4b) were in good agreement with the CV data in determining the number of electrons involved in the first oxidation wave, but at the slower scan rate of the UME CV data, the second TTF oxidation wave became irreversible, suggesting some decomposition of the TTF dication species in structure **34**.

The cyclic voltammogram and the differential pulse voltammogram of the stable system **35**, bearing eight TTF substituents in acetonitrile/dichloromethane in the presence of 2,3-dichloronaphthoquinone, showed a relative intensity ratio of 6 ± 1 between the anodic and cathodic scans, which is slightly lower than the theoretical ratio of 8 for simultaneous oxidation of all the TTF units. Nonetheless, these data imply that the TTF groups were not interacting with one another (Figure 4). This is a similar situation to the multi-ferrocene and iron-sandwich derivatives **3**, **6**, **12**, **15**, and **16**.

Chemical oxidation of the TTF groups in compounds **34** and **35** has been achieved by reaction with an excess of iodine in dichloromethane solution, leading to new low-energy absorptions in the UV/visible spectra which are diagnostic of TTF cation radicals:[44,45] the broad absorption at $\lambda_{max} = 830$ nm for the iodide salt of **35** suggests the formation of aggregated TTF species. A charge transfer complex formed by **35** and tetracyano-p-quinodimethane (TCNQ) has been isolated as an insoluble black powder. The stoichiometry is $(\mathbf{35})_1$:(TCNQ)$_3$ (i.e. 8 TTF units:3

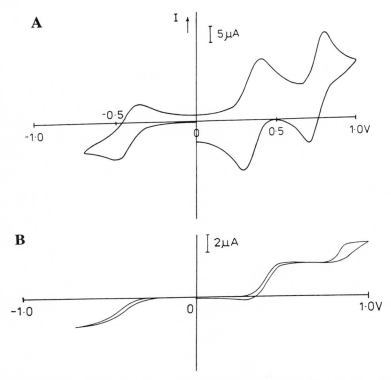

Figure 4. Electrochemistry of dendrimer **35**: (**a**) classical CV, in the presence of 2,3-dichloronaphthoquinone as internal standard, which gives rise to the wave at negative potential (solvent MeCN, electrolyte Bu$_4$N$^+$ PF$_6^-$, Pt electrode, versus SCE, scan rate 100 mV s^{-1}). (**b**) Ultramicroelectrode CV, in the presence of 2,3-dichloro-naphthoquinone [solvent MeCN/CH$_2$Cl$_2$ (1:1 v/v) electrolyte Bu$_4$N$^+$ PF$_6^-$, Pt electrode, versus SCE, scan rate 50 mV s^{-1}].

TCNQ units) as judged by elemental analysis, and the complex is a semiconductor, $\sigma_{rt} = 2.2 \times 10^{-3}$ Scm^{-1} (two-probe measurement on compressed pellets.) The solid state IR spectrum of the complex exhibited a single nitrile absorption peak at 2180 cm^{-1}, which suggests that each of the TCNQ molecules is present as the radical anion. The stoichiometry of the complex dictates that the dendrimer is partially oxidized, and, although the detailed structure is not known, it seems likely that it is the long-range (inter-dendrimer) interaction of TTF radical cations rather than the TCNQ radical anions which gives rise to the observed conductivity. This family of materials affords unprecedented opportunities for the study of charge transfer interactions involving dendrimers and related hyperbranched systems.

Tetrathio–TTF derivatives are relatively easy to prepare, and in collaboration with Becher's group, we have constructed the pentakis-TTF derivative **38** by

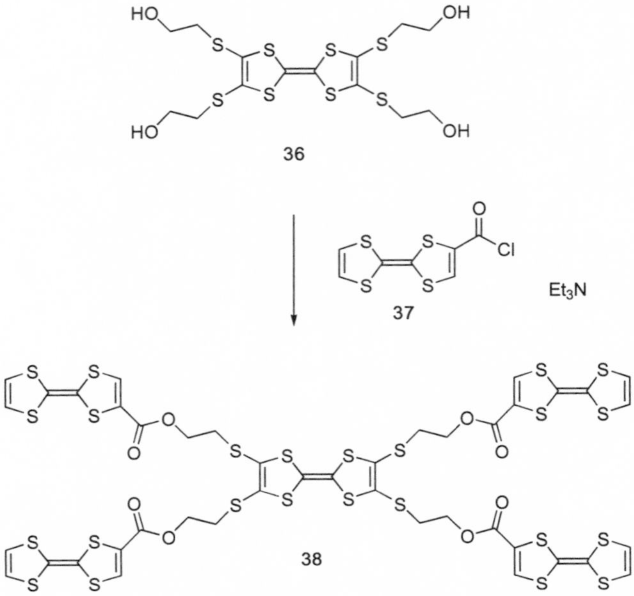

Scheme 8.

reaction of tetra-ol **36** with TTF–carbonyl chloride **37** under basic conditions (Scheme 8).[47] Compound **38** is a model for redox-active dendrimers with substituents sprouting out from a tetrafunctional TTF core. Compound **38** shows three quasi-reversible redox couples in the cyclic voltammogram (Figure 5). The first couple, $E_1^{1/2} = +0.60$ V (4 electrons) (vs. SCE), corresponds to the first oxidation of the four peripheral TTFs to form a tetracation; the first oxidation of the central TTF with loss of one electron gives rise to the second, poorly resolved couple, $E_2^{1/2} = +0.73$ (1 electron) (it is known that thioalkyl substitution raises the oxidation potential of TTF)[48] and finally, the second oxidation of the core TTF and the four peripheral TTFs appear to coincide to give the third redox couple, $E_3^{1/2} = +0.78$ (5 electrons) generating a species bearing 10 positive charges.

The related pentakis-TTF derivatives **41**, synthesized by Becher et al. by reaction of tetraiodide **39** with four equivalents of the thiolate anion liberated upon treatment of **40** with cesium hydroxide (Scheme 9), displayed qualitatively similar redox behavior to **38** with noninteracting TTF units.[42b] The attachment of branched multiredox wedges (e.g. **32**) to the tetrathio-TTF core should provide interesting systems in which a central TTF unit may be shielded by the outer layers and should oxidize more readily.

Jørgensen et al. have synthesized the bis(arborol)-TTF derivative **46** (presumed to be a mixture of 4,4'- and 4,5'-isomers, with respect to the position of substituents

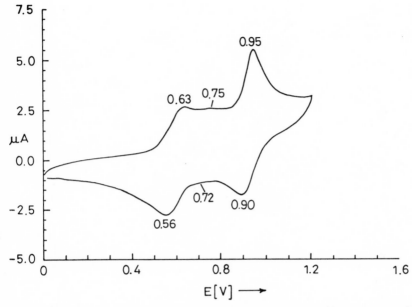

Figure 5. Cyclic voltammogram of pentakis-TTF derivative **38** (solvent MeCN, electrolyte Bu_4N^+ PF_6^-, Pt electrode, versus SCE) (redrawn from Ref. 47).

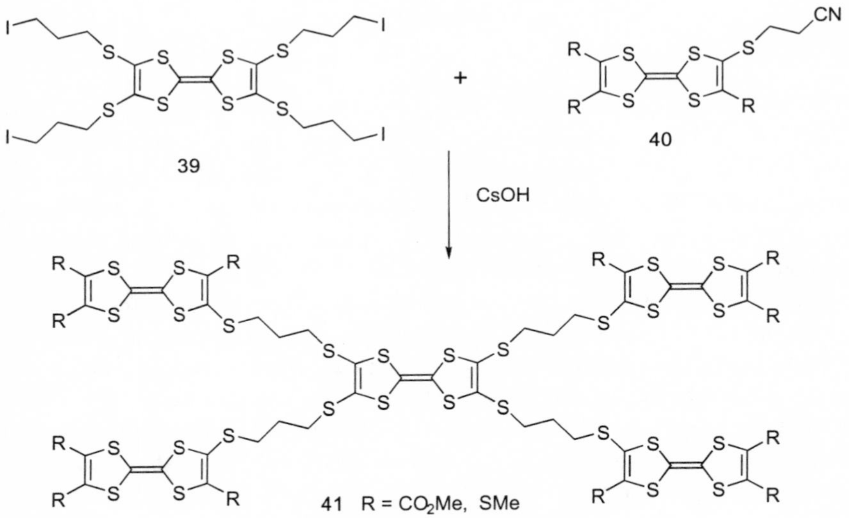

Scheme 9.

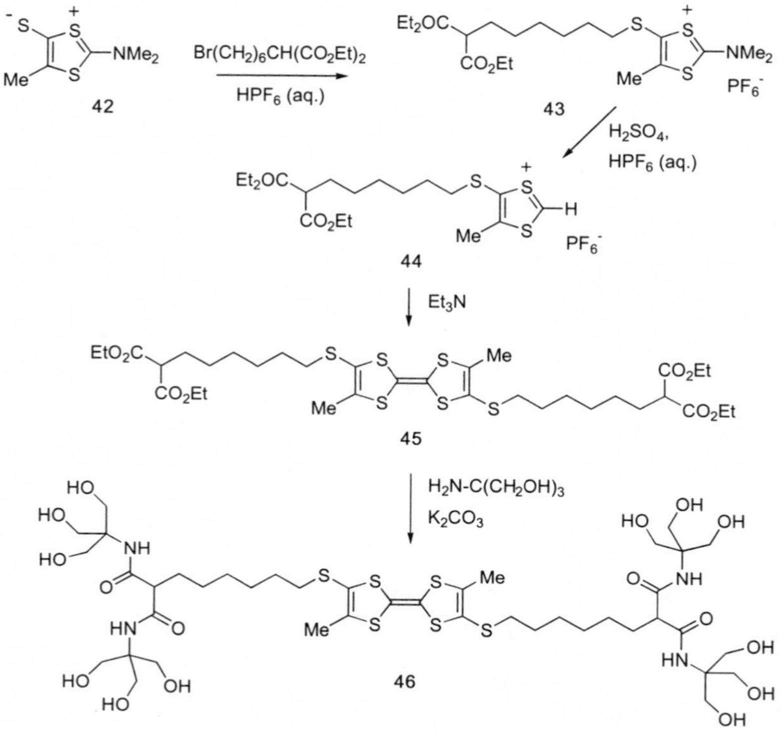

Scheme 10.

on the TTF ring) using the methodology in Scheme 10, with the ingenious aim of constructing a self-assembling molecular wire which could be a conductor along an internal core of stacked, partially-oxidized TTFs surrounded by insulating arborol groups.[49] The mesoionic thiolate **42** was alkylated to yield **43**, which was converted via 1,3-dithiolium cation salts **43** and **44** into TTF derivative **45**. The arboreal groups were then attached to **45** to afford **46**. Cyclic voltammetry measurements established that the redox potentials of the TTF unit in **46** were not altered significantly by the substituents. The compound formed a gel comprising thin string-like aggregates with lengths on the order of microns and diameters of 30–100 nm, as revealed by optical and atomic force microscopy. Spectroscopy of I_2-doped **46**, both as a solution of the gel, and as a dried gel, showed a low-energy absorption at $\lambda_{max} = 874$ nm, suggesting that the oxidized TTF units formed stacks, although no conductivity data were reported. Molecular modeling of **46** suggested that the *cis*-isomer would form a helical structure, whereas the *trans*-isomer would form a stacked structure. Calculations performed on a dimer of two *trans*-isomers showed

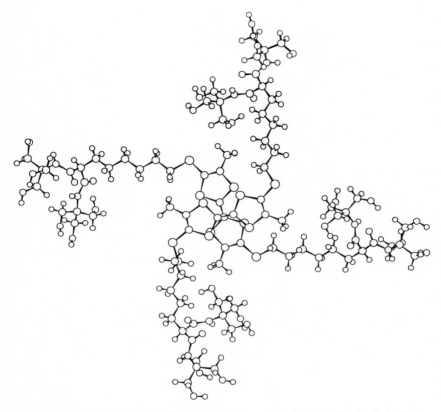

Figure 6. Molecular modelling of *trans*-**46** as a stack (from data in Ref. 49).

that the planes of the two TTF units were parallel with and orthogonal to each other (Figure 6). Repetition of this dimer motif would give a stack with a diameter of ca. 3.5 nm.

D. Porphyrin and Phthalocyanine Derivatives

Dendritic derivatives of these macrocycles can be placed in the wider context of studies on metalloporphyrins with sterically hindered faces which have been designed in attempts to mimic the properties of heme proteins and chlorophylls, and there are suggestions that steric isolation of the metalloporphyrin nucleus is important in certain biological functions.[50] The redox properties of metalloporphyrins are well-documented; they are dominated by two, reversible one-electron transfers involving both the metal and the ligand.[51] The first dendritic porphyrins of general structure **47** and their Zn complexes were reported by Inoue et al.[52] who

47   Ar = 1,3,5-C$_6$H$_3$

used tetrakis(3′,5′-dihydroxyphenyl)porphine as a core unit, which by a divergent route became embedded within poly(arylether) branches of the type pioneered by Hawker and Fréchet.[53] Based upon ^{1}H NMR data it was proposed that the dendrimers **47** were nearly globular in shape. Fluorescence quenching studies suggested that the steric bulk of the higher generation dendrimers **47** prevented access of a large quencher molecule to the metalloporphyrin core, but behaved as a trap for a small quencher molecule. Electrochemical data for **47** were not reported.[52]

Electrochemical studies on dendritic Zn– and Fe–porphyrins (as the FeCl system) have been reported by Diederich et al.[54] Tetra-acid **48** is a key building block, which was derivatized with polyether–amide substituents to furnish hyperbranched systems with multifunctionality, such as compound **49**, and by further modification the water-soluble glycol derivatives **50** were obtained (Scheme 11). From a series of cyclic voltammetric and spectroelectrochemical studies on **50** and analogs, it was demonstrated that for higher generation derivatives the pendant arms play a significant role in modulating the redox properties of the metalloporphyrin cores. In particular, the data showed that microenvironmental polarity strongly influenced the redox potential of electrochemical reactions at the Fe–porphyrin core. The potential of the Fe^{2+}/Fe^{3+} couple in water was raised by ca. 300 mV in the higher generation derivatives compared to lower generation systems. This is consistent with the relatively open dendritic branches of lower generation systems allowing solvation of the core, whereas the more dense packing of higher generations impedes solvation, which in turn destabilizes the Fe^{3+} state. Materials of this family were recognized as models of electron transfer proteins like cytochrome *c*, for which the oxidation potential of the Fe^{2+}/Fe^{3+} couple in aqueous solution is 300–400 mV more positive than similar heme proteins which lack the hydrophobic peptide shell.[54]

To the best of our knowledge dendritic phthalocyanines have not yet been reported. The phthalocyanine unit could act as a very interesting octadirectional core, with rich electrochemical, spectroscopic, and photophysical properties.[55] A prototype system **51**, which may point a way forward to novel macromolecules incorporating different multiredox sites, has recently been synthesized in our laboratory starting from TTF derivative **31**.[56] The cyclic voltammetric and spectroelectrochemical properties indicated that the peripheral TTF groups did not interact electronically with the phthalocyanine core, and the data were consistent with aggregation in solution, although studies were hampered by the insolubility of oxidized states (viz. TTF^{+} and TTF^{2+} species). Analogs with increased solubility are currently under investigation in our laboratory.

E. Fullerenes

One of the most important properties of C_{60} is its ability to accept sequentially up to six electrons.[57] Polymer-modified fullerenes may lead to novel processable materials. Hawker and co-workers prepared the monosubstituted C_{60} azafulleroid **52** possessing the expected ring-opened annulene structure by reaction of the corresponding dendritic benzyl azide with C_{60}.[58] Compound **52** was found to be very soluble in organic solvents (unlike C_{60}) and the cyclic voltammogram in dichloromethane revealed three clearly resolved reduction waves at $E^{1/2} = -0.65$, -1.05, and -1.50 V (vs. Ag/AgCl). Multiaddition to C_{60}, which often leads to complex product mixtures,[59] was retarded in this system, possibly due to steric hindrance around the C_{60} nucleus. Dendritic encasement may provide a fruitful method for isolating the C_{60} core, and thereby modifying its physical properties compared to free fullerenes in bulk solid or liquid phase.

Previously Fréchet et al. had prepared a dendritic methanofullerene, but its redox properties were not reported.[60]

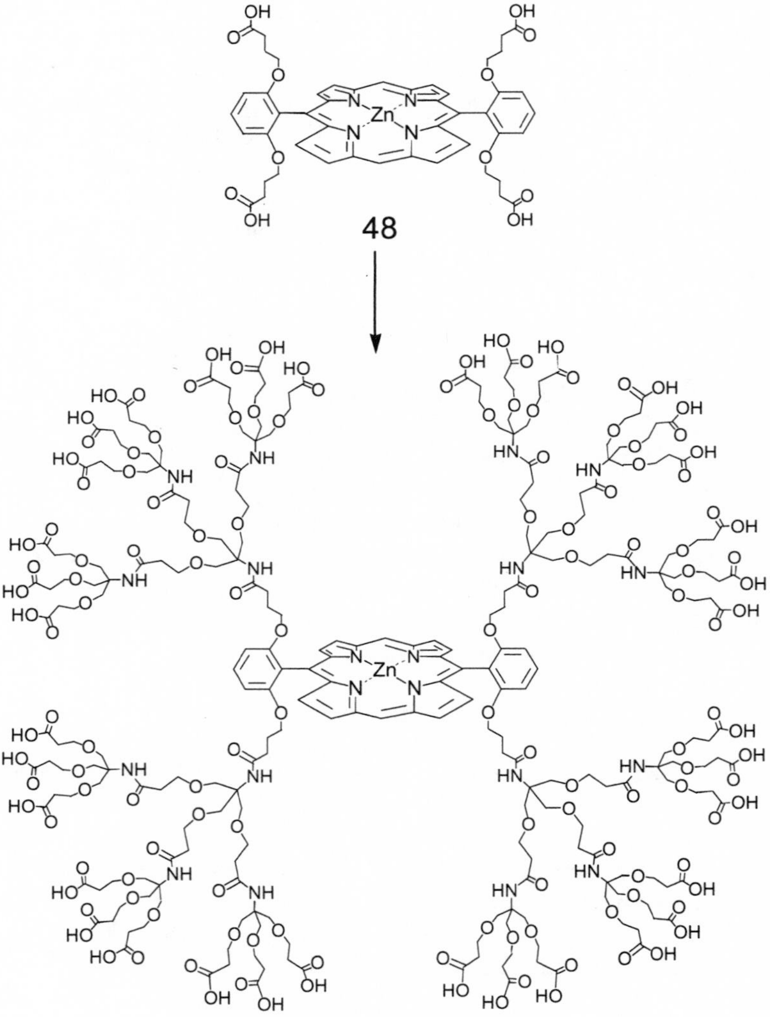

48

49   *(continued)*

Scheme 11.

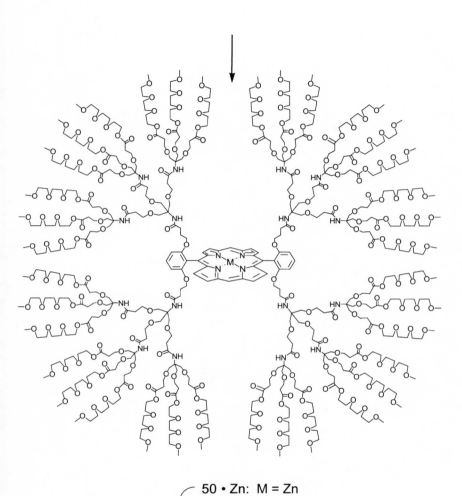

50 • Zn: M = Zn

51 • H$_2$: M = H$_2$

52 • FeCl: M = FeCl

Scheme 11. (Continued)

51

52

F. Poly(amidoamine) Derivatives

Miller et al.[61] have synthesized poly(amidoamine) (PAMAM) dendrimers[62] modified at the termini of the branches with redox-active imide groups, e.g. first-generation system **54** containing 12 peripheral groups (Scheme 12). These dendrimers were synthesized by reaction of the anhydride group of the mono-imide reagent **53** with the preformed PAMAM dendrimer (generations 1–6); assuming complete surface functionalization the sixth-generation product would possess 196 imide groups. It was concluded that loading of the imide groups on the polymer surface was ca. 100% for all the generations studied, based on UV spectroscopic and electrochemical coulometric data. Electrochemical reduction in both DMF and in water led to the sequential formation of anion radical and dianion species at potentials E^{red} of between −0.2 and −0.9 V, (vs. SCE). The data, especially the breadth of the first redox wave, suggested π-aggregation or stacking of the imide anion radicals, although it is not clear if these interactions are intra- or/and intermolecular. These systems contrast, therefore, with the noninteracting peripheral ferrocene or TTF groups in the dendrimers discussed above.

Meijer and co-workers have synthesized a fifth-generation poly(propylene imine) dendrimer with 64-peripheral *N-t*-butoxycarbonyl-protected L-phenylalanine groups. Various guest molecules have been trapped within the internal cavities, including TCNQ. UV/visible and ESR spectroscopic data indicated that electron transfer occurs within the dendritic "box" to form the TCNQ radical anion.[14b,63] Cyclic voltammetry of the dendrimer shows an irreversible oxidation at ca. 0.85 V (vs. SCE) in dichloromethane, and a photoinduced electron transfer reaction to C_{60} occurred in solution. This reaction exhibited an interesting size effect: persistent photoinduced electron transfer occurred only for dendrimer generations $G \geq 2$, demonstrating that new properties of dendrimers emerge at higher generations.[64]

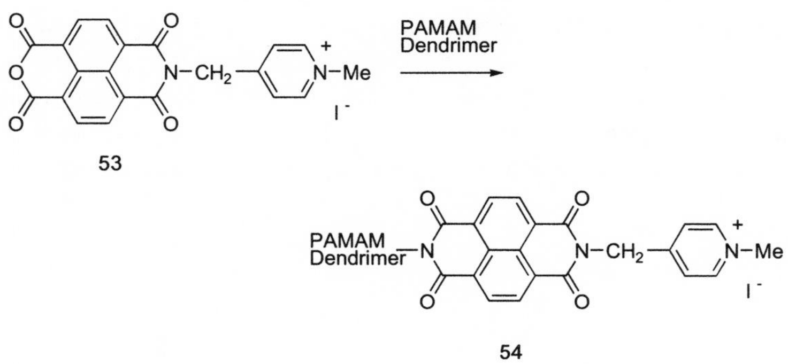

53

54

Scheme 12.

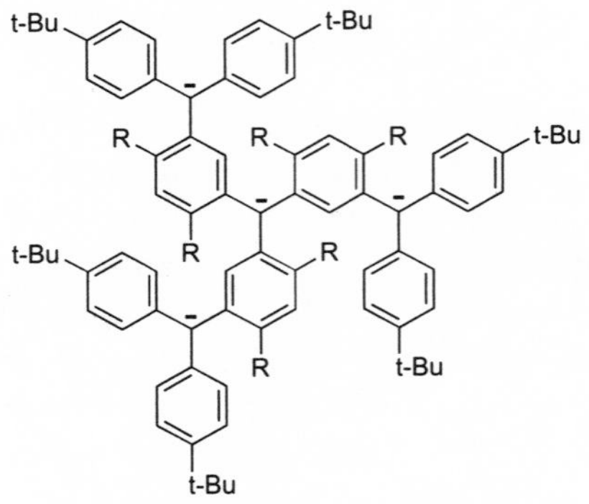

55 R = H, alkyl

G. Poly(arylmethane) Systems

Rajca and co-workers have studied star-branched and dendritic high-spin polyradicals which are potential organic magnets.[65] Representative data were obtained for the model tetra-anionic compound **55**. Three redox waves were observed by cyclic voltammetry and differential pulse voltammetry for a four-electron process between the potentials of -2.00 and -1.20 V (vs. SCE). Electrochemical experiments with these materials have usually been performed at 200 K. The polyradicals, which are less stable for systems with more unpaired electrons, have been characterized by spectroscopic studies, ESR data, and SQUID magnetometry.

H. Cobalt(salen) Analogs

The Co(II)salen system **56** [salen = N,N-bis(salicylidene)ethylenediamine)] is known to complex oxygen reversibly, to undergo a quasi-reversible one-electron redox reaction (the Co^{2+}/Co^{3+} couple), and to possess interesting catalytic activity, providing a model for enzymes with monooxygenase, dioxygenase, and peroxidase activity.[66] Moors and Vögtle have used a tris(2-aminoethyl)amine core to synthesize the analogous tri-cobalt(II) complex **57**,[67] cyclic voltammetry of which showed reversible redox behavior and increased stability in air compared to salen complex **56**.[8b]

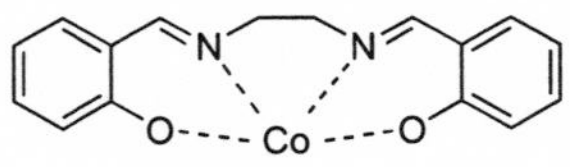

56

57

IV. CONCLUSIONS AND PROSPECTS

Dendrimers and hyperbranched molecules provide unique molecular scaffolds for the emplacement of multiples of functional groups in predetermined spatial arrangements. Redox activity is a property of these systems which is now attracting widespread attention. The electrochemistry of redox groups and the stability of charged species embedded at the core of a dendrimer can be modulated by the surrounding microenvironment of the substituent arms (e.g. porphyrin **50**). Peripheral substituents can undergo reversible, multielectron transfer waves (e.g. ferrocene and TTF derivatives), and prototypes of new electronically conducting complexes (**35**-TCNQ) and π-electron delocalized systems (**54**) have been reported. There is scope for more detailed studies on the kinetics of electron transfer processes at the surfaces and in the interiors of these dendrimers. Analysis of the electrochemical behavior of macromolecules which contain several electroactive sites is rarely straightforward. The potentials of the half-reactions of successive electron transfers in such molecules are dependent upon several factors, such as the extent of interaction between the sites, solvation of charged species, ion pairing, and intramolecular structural changes. For polymers containing identical, noninteracting centers, Bard et al. have demonstrated that the successive electron transfers can follow simple statistics, although complications can arise due to adsorption, precipitation, and slow electron transfer kinetics.[68] Thin-layer cyclic voltammetry (TLCV) may prove to be a valuable technique for studying the electrochemistry of redox-active dendrimers, since, contrary to conventional cyclic voltammetry, in TLCV the current is not limited by the kinetics of mass transfer to the electrode.[69]

Bard has suggested that the multiple redox properties of these materials offers special potential in the field of multielectron catalysis, including such processes as the reduction of dioxygen to water, or dinitrogen to ammonia, under ambient conditions in ways which mimic enzymatic processes.[28] Interesting target materials, not yet reported in the literature, are segment-block dendritic copolymers which possess wedges containing strong electron-donor groups, e.g. ferrocene or tetrathiafulvalene, and wedges which contain strong electron-acceptor groups, e.g. quinones, tricyanovinyl, or pyridinium groups, at the chain ends in opposing segments. The convergent methodology of Hawker and Fréchet[70] should be appropriate for these syntheses once the redox wedges have been grown. High dipole moments, optical nonlinearity, and vectoral intramolecular electron transport are potentially very interesting properties of such systems. The possibility of channeling light energy or electrons in a well-defined manner holds promise for the development of nanoscale devices for light harvesting and photoconversion. Dendrimer arms attached to a redox-active group may be useful as "stoppers" in rotaxane structures of the type studied by Stoddart et al.,[71] and other intertwined molecules[71b] which are assembled and stabilized by virtue of charge transfer interactions.

Some of the materials highlighted in this review offer novel redox-active cavities, which are candidates for studies on chemistry within cavities, especially processes which involve molecular recognition by donor–acceptor π–π interactions, or by electron transfer mechanisms, e.g. coordination of a lone pair to a metal center, or formation of radical cation/radical anion pairs by charge transfer. The attachment of redox-active dendrimers to electrode surfaces (by chemical bonding, physical deposition, or screen printing) to form modified electrodes should provide interesting novel electron relay systems.

ACKNOWLEDGMENTS

We thank the U.K. Engineering and Physical Sciences Research Council for funding our work on dendrimers, and Dr. A. J. Moore and Dr. M. Blower for their contributions to this project. M. R. B. gratefully acknowledges the University of Durham for the award of a Sir Derman Christopherson Research Fellowship, and Ciba-Geigy for a Fellowship enabling him to visit Risø National Laboratory, Roskilde, and the University of Copenhagen. We thank Professor F. Diederich for the structures in Scheme 11, Dr. M. Jørgensen for Figure 6, and Professor F. Vögtle and Professor E. Meijer for preprints.

REFERENCES AND NOTES

1. Buhleier, E.; Wehner, W.; Vögtle, F. *Synthesis* **1978**, 155.
2. (a) Newkome, G. R.; Moorefield, C. N.; Baker, G. R. *Aldrichim. Acta* **1992**, *25*, 31; (b) Moorefield, C. N.; Newkome, G. R. In *Advances in Dendritic Macromolecules*; Newkome, G. R., Ed.; JAI Press: London, 1994, Volume 1; (c) Newkome, G. E.; Moorefield, C. N. In *Mesomolecules: From Molecules to Materials;* Mendenhall, G. D.; Greenberg, A.; Liebman, J. F., Eds.; Chapman and Hall: New York, 1995, 27.
3. Fréchet, J. M. J. *Science* **1994**, *263*, 1710.

4. (a) Tomalia, D. A.; Durst, H. D. In *Supramolecular Chemistry I - Directed Synthesis and Molecular Recognition*; Weber, E., Ed.; Springer-Verlag: Berlin, 1993, 193; (b) Tomalia, D. A.; Naylor, A. M.; Goddard III, W. A. *Angew. Chem. Int. Ed. Engl.* **1990**, *29*, 138; (c) Tomalia, D. A. *Adv. Mater.* **1994**, *6*, 529.

5. (a) Dagani, R. *Chem. and Eng. News* **1993**, February 1, 28; (b) Service, R. F. *Science* **1995**, *267*, 458; (c) Tomalia, D. A.; Dvornic, P. R. *Nature (London)* **1994**, *372*, 617; (d) Ardoin, N.; Astruc, D. *Bull. Soc. Chim. Fr.* **1995**, *132*, 875.

6. Kuzdzal, S. A.; Monnig, C. A.; Newkome, G. R.; Moorefield, C. N. *J. Chem. Soc., Chem. Commun.* **1994**, 2139.

7. Xu, Z.; Moore, J. S. *Angew. Chem. Int. Ed. Engl.* **1993**, *32*, 1354.

8. For reviews which emphasize the functional properties of dendrimers see: (a) Issberner, J.; Moors, R.; Vögtle, F. *Angew. Chem. Int. Ed. Engl.* **1994**, *33*, 2413; (b) Moors, R.; Vögtle, F. In *Advances in Dendritic Macromolecules*; Newkome, G. R., Ed.; JAI Press: London, 1996, Volume 2, p. 41.

9. (a) Nagasaki, T.; Ukon, M.; Arimori, S.; Shinkai, S. *J. Chem. Soc., Chem. Commun.* **1992**, 608; (b) Nagasaki, T.; Kimura, O.; Ukon, M.; Arimori, S.; Hamachi, I.; Shinkai, S. *J. Chem. Soc., Perkin Trans. 1* **1994**, 75.

10. (a) Newkome, G. R.; Liu, X.; Weis, C. D. *Tetrahedron: Asymmetry* **1991**, *2*, 957; (b) Chow, H-F.; Mak, C. C. *J. Chem. Soc., Perkin Trans. 1* **1994**, 2223; (c) Seebach, D.; Lapierre, J. M.; Greiveldinger, G.; Skobridis, K. *Helv. Chim. Acta* **1994**, *77*, 1673; (d) Kremers, J. A.; Meijer, E. W. *J. Org. Chem.* **1994**, *59*, 4262; (e) Brunner, H.; Bublak, P. *Synthesis* **1995**, 36; (f) Jansen, J. F. G. A.; Peerlings, H. W. I.; de Brabander-Van den Berg, E. M. M.; Meijer, E. W. *Angew. Chem. Int. Ed. Engl.* **1995**, *34*, 1206.

11. Percec, V.; Chu, P.; Kawasumi, M. *Macromolecules* **1994**, *27*, 4441.

12. (a) Lee, J-J.; Ford, W. T.; Moore, J. A.; Li, Y. *Macromolecules* **1994**, *27*, 4632; (b) van der Made, A. W.; van Leeuwan, P. W. N. M.; de Wilde, J. C.; Brander, R. A. C. *Adv. Mater.* **1993**, *5*, 466.

13. Campagne, S.; Denti, G.; Serroni, S.; Juris, A.; Venturi, M.; Ricevuto, V.; Balzani, V. *Chem. Eur. J.* **1995**, *1*, 211.

14. (a) Ottaviani, M. F.; Bossmann, S.; Turro, N. J.; Tomalia, D. A. *J. Am. Chem. Soc.* **1994**, *116*, 661; (b) Jensen, J. F. G. A.; de Brabander-Van den Berg, E. M. M.; Meijer, E. W. *Science* **1994**, *266*, 1226; (c) Mattei, S.; Seiler, P.; Diederich, F.; Gramlich, V. *Helv. Chim. Acta* **1995**, *78*, 1904.

15. Turro, N. J.; Barton, J. K.; Tomalia, D. A. *Accts. Chem. Res.* **1991**, *24*, 332.

16. (a) *Ferrocenes*; Togni, A.; Hayashi, T., Eds.; VCH: Weinheim, Germany; (b) Foucher, D. A.; Ziembinski, R.; Rulkens, R.; Nelson, J.; Manners, I. In *Inorganic and Organometallic Polymers II: Advanced Materials and Intermediates*; Wisian-Neilson, P.; Allcock, H. R.; Wynne, K. J., Eds.; ACS Symposium Series 572, ACS, Washington, DC, 1994, Chapter 33.

17. (a) Flanagan, J. B.; Margel, S.; Bard, A. J.; Anson, F. C. *J. Amer. Chem. Soc.* **1978**, *100*, 4248; (b) D'Silva, C.; Afeworki, S.; Parri, O. L.; Baker, P. K.; Underhill, A. E. *J. Mater. Chem.* **1992**, *2*, 225; and references therein.

18. Hale, P. D.; Inagaki, T.; Karan, H. I.; Okamoto, Y.; Skotheim, T. A. *J. Amer. Chem. Soc.* **1989**, *111*, 3482.

19. Wright, M. E.; Cochran, B. B.; Toplikar, E. G.; Lackritz, H. S.; Kerney, J. T., in Ref. 16b, Chapter 34.

20. (a) Miller, J. S.; Epstein, A. J. *Angew. Chem. Int. Ed. Engl.* **1994**, *33*, 385; (b) Gatteshi, D. *Adv. Mater.* **1994**, *6*, 635.

21. Chaofeng, Z.; Wrighton, M. S. *J. Amer. Chem. Soc.* **1990**, *112*, 7578.

22. Casado, C. M.; Cuadrado, I.; Morán, M.; Alonso, B.; Lobete, F.; Losada, J. *Organometallics* **1995**, *14*, 2618.

23. Casaco, C. M.; Cuadrado, I.; Morán, M.; Losada, J. *J. Chem. Soc., Chem. Commun.* **1994**, 2575.

24. Astruc, D. *New J. Chem.* **1992**, *16*, 305.

25. Moulines, F.; Djakovitch, L.; Boese, R.; Gloaguen, B.; Thiel, W.; Fillaut, J-L.; Delville, M-H.; Astruc, D. *Angew. Chem. Int. Ed. Engl.* **1993**, *32*, 1075.

26. Fillaut, J-L.; Linars, J.; Astruc, D. *Angew. Chem. Int. Ed. Engl.* **1994**, *33*, 2460.

27. Astruc, D. *Accts. Chem. Res.* **1986**, *19*, 377.

28. Bard, A. J. *Nature* **1995**, *374*, 13.

29. Polymer electrodes modified with pendant metal-bipyridyl groups have been studied in detail: Eaves, J. G.; Munro, H. S.; Parker, D. *Inorg. Chem.* **1987**, *26*, 644; and references therein.

30. (a) Juris, A.; Balzani, V.; Barigelletti, F.; Campagne, S.; Belser, P.; von Zelewsky, A. *Coord. Chem. Rev.* **1988**, *84*, 85; (b) Meyer, T. J. *Accts. Chem. Res.* **1989**, *22*, 163; (c) Roffia, S.; Marcaccio, M.; Paradisi, C.; Paolucci, F.; Balzani, V.; Denti, G.; Serroni, S.; Campagna, S. *Inorg. Chem.* **1993**, *32*, 3003; (d) Denti, G.; Campagne, S.; Balzani, V. In *Mesomolecules: From Molecules to Materials*; Mendenhall, G. D.; Greenberg, A.; Liebman, J. F., Eds.; Chapman and Hall: New York, 1995, 69.

31. Serroni, S.; Denti, G.; Campagne, S.; Juris, A.; Ciano, M.; Balzani, V. *Angew. Chem. Int. Ed. Engl.* **1992**, *31*, 1493.

32. Newkome, G. Quoted in Ref. 5a. For a recent overview of inorganic and organic materials with "open" frameworks see: Bowes, C. L.; Ozin, G. A. *Adv. Mater.* **1996**, *8*, 13.

33. Newkome, G. R.; Cardullo, F.; Constable, E. C.; Moorefield, C. N.; Cargill Thompson, A. M. W. *J. Chem. Soc., Chem. Commun.* **1993**, 925.

34. Newkome, G. R.; Güther, R.; Moorefield, C. N.; Cardullo, F.; Echegoyen, L.; Pérez-Cordero, E.; Luftmann, H. *Angew. Chem. Int. Ed. Engl.* **1995**, *34*, 2023.

35. (a) Constable, E. C.; Harvesson, P. *Chem. Commun.* **1996**, 33; (b) Constable, E. C.; Cargill Thompson, A. M. W.; Harveson, P.; Macko, L.; Zehnder, M. *Chem. Eur. J.* **1995**, *1*, 360.

36. Achar, S.; Puddephatt, R. J. *J. Chem. Soc., Chem. Commun.* **1994**, 1895.

37. Achar, S.; Puddephatt, R. J. *Angew. Chem. Int. Ed. Engl.* **1994**, *33*, 847.

38. (a) Bryce, M. R. *Chem. Soc. Rev.* **1991**, *20*, 355; (b) *J. Mater. Chem.* **1995**, *5*, 1469–1760 (Special Issue on Molecular Conductors).

39. Review: Adam, M.; Müllen, K. *Adv. Mater.* **1994**, *6*, 439.

40. Frenzel, S.; Arndt, S.; Gregorious, R. Ma.; Müllen, K. *J. Mater. Chem.* **1995**, *5*, 1529.

41. Garín, J. *Adv. Heterocycl. Chem.* **1995**, *62*, 249.

42. (a) Garín, J.; Orduna, J.; Uriel, S.; Moore, A. J.; Bryce, M. R.; Wegener, S.; Yufit, D. S.; Howard, J. A. K. *Synthesis* **1994**, 489; (b) Lau, J.; Simonsen, O.; Becher, J. *Synthesis* **1995**, 521.

43. Jørgensen, T.; Hansen, T. K.; Becher, J. *Chem. Soc. Rev.* **1994**, *23*, 41.

44. Bryce, M. R.; Devonport, W.; Moore, A. J. *Angew. Chem. Int. Ed. Engl.* **1994**, *33*, 1761.

45. Bryce, M. R.; Devonport, W. *Proceedings of the European Materials Research Society Meeting*, Strasbourg, May 1995, *Synth. Metals* **1996**, *76*, 305.

46. Devonport, W. Ph.D. Thesis, University of Durham, **1995**.

47. Marshallsay, G. J.; Hansen, T. K.; Moore, A. J.; Bryce, M. R.; Becher, J. *Synthesis* **1994**, 926.

48. Saito, G. *Pure Appl. Chem.* **1987**, *59*, 999.

49. Jørgensen, M.; Bechgaard, K.; Bjørnholm, T.; Sommer-Larsen, P.; Hansen, L. G.; Schaumburg, K. *J. Org. Chem.* **1994**, *59*, 5877.

50. Bonar-Law, R. P.; Mackay, L. G.; Sanders, J. M. K. *J. Chem. Soc., Chem. Commun.* **1993**, 456.

51. Fuhrhop, J.-H.; Kadish, K. M.; Davis, D. G. *J. Amer. Chem. Soc.* **1973**, *95*, 5140.

52. Jin, R-H.; Aida, T.; Inoue, S. *J. Chem. Soc., Chem. Commun.* **1993**, 1260.

53. (a) Hawker, C.; Fréchet, J. M. J. *J. Chem. Soc., Chem. Commun.* **1990**, 1010; (b) Wooley, K. L.; Hawker, C. J.; Fréchet, J. M. J. *Angew. Chem. Int. Ed. Engl.* **1994**, *33*, 82.

54. (a) Dandliker, P. J.; Diederich, F.; Gross, M.; Knobler, C. B.; Louati, A.; Sanford, E. M. *Angew. Chem. Int. Ed. Engl.* **1993**, *33*, 1739; (b) Dandliker, P. J.; Diederich, F.; Gisselbrecht, J-P.; Louati, A.; Gross, M. *Angew. Chem. Int. Ed. Engl.* **1994**, *34*, 2725.

55. Leznoff, C. C.; Lever, A. B. P., Eds.; *Phthalocyanines, Properties and Applications*; VCH: Weinheim, 1989–1993, Vols. 1–3.

56. Blower, M. A.; Bryce, M. R.; Devonport, W. *Adv. Mater.* **1996**, *8*, 63.

57. (a) Dubois, D.; Moninot, G.; Kutner, W.; Jones, M. T.; Kadish, K. *J. Phys. Chem.* **1992**, *96*, 7137; (b) Xie, Q.; Pérez-Cordero, E.; Echegoyen, L. *J. Am. Chem. Soc.* **1992**, *114*, 3978.

58. Hawker, C. J.; Wooley, K. L.; Fréchet, J. M. J. *J. Chem. Soc., Chem. Commun.* **1994**, 925.
59. (a) Hirsch, A. *The Chemistry of Fullerenes*; Thieme: Stuttgart, 1994; (b) Diederich, F.; Thilgen, C. *Science* **1996**, *271*, 317.
60. Wooley, K. L.; Hawker, C. J.; Fréchet, J. M. J.; Wudl, F.; Srdanov, G.; Shi, S.; Li, C.; Kao, M. *J. Amer. Chem. Soc.* **1993**, *115*, 9836.
61. Miller, L. L.; Hashimoto, T.; Tabakovic, I.; Swanson, D. R.; Tomalia, D. A. *Chem. Mater.* **1995**, *7*, 9.
62. For pioneering work on PAMAM dendrimers see: Tomalia, D. A.; Baker, H.; Dewald, J. R.; Hall, M.; Kallos, G.; Martin, S.; Roeck, J.; Ryder, J.; Smith, P. *Macromolecules* **1986**, *19*, 2466.
63. Bosman, A. W.; Jansen, J. F. G. A.; Janssen, R. A. J.; Meijer, E. W. *ACS Polym. Mat. Sci. and Eng.* **1995**, *73*, 340.
64. Janssen, R. A. J.; Jansen, J. F. G. A.; van Haare, J. A. E. H.; Meijer, E. W. *Adv. Mater.* **1996**, *8*, 494.
65. (a) Rajca, A. In Ref. 2b, p. 113; (b) Utamapanya, S.; Rajca, A. *J. Amer. Chem. Soc.* **1991**, *113*, 9242.
66. Eichhorn, E.; Rieker, A.; Speiser, B. *Angew. Chem. Int. Ed. Engl.* **1992**, *31*, 1215.
67. Moors, R.; Vögtle, F. *Chem. Ber.* **1993**, *126*, 2133.
68. Flanagan, J. B.; Margel, S.; Bard, A. J.; Anson, F. C. *J. Amer. Chem. Soc.* **1978**, *100*, 4248.
69. Hubbard, A. T.; Anson, F. C. In *Electroanalytical Chemistry*; Bard, A. J., Ed.; Marcel Dekker, New York, 1970, Vol. 4, pp. 129–210.
70. (a) Hawker, C. J.; Fréchet, J. M. J. *J. Am. Chem. Soc.* **1992**, *114*, 8405; (b) Wooley, K. L.; Hawker, C. J.; Fréchet, J. M. J. *J. Am. Chem. Soc.* **1993**, *115*, 11496.
71. (a) Bissel, R. A.; Córdova, E.; Kaifer, A. E.; Stoddart, J. F. *Nature* **1994**, *369*, 133; (b) Amabilino, D. B.; Stoddart, J. F. *Chem. Rev.* **1995**, *95*, 2725.

ORGANOMETALLIC DENDRITIC MACROMOLECULES:
ORGANOSILICON AND ORGANOMETALLIC ENTITIES AS CORES OR BUILDING BLOCKS

Isabel Cuadrado, Moisés Morán, José Losada,
Carmen M. Casado, Carmen Pascual,
Beatriz Alonso, and Francisco Lobete

Advances in Dendritic Macromolecules
Volume 3, pages 151–195
Copyright © 1996 by JAI Press Inc.
All rights of reproduction in any form reserved.
ISBN: 0-7623-0069-8

ABSTRACT

This review summarizes pioneering relevant works of expert groups on organomet-allic dendritic macromolecules, together with our initial efforts in this developing field. Throughout the chapter, the synthetic aspects are especially emphasized. Likewise, we describe in detail the redox properties exhibited by a new family of silicon-based ferrocenyl-containing dendrimers and we also mention some of their relevant applications.

I. INTRODUCTION

The design and synthesis of well-defined dendritic macromolecules, characterized by their highly branched three-dimensional architectures, and exhibiting a unique combination of new chemical and physical properties, is a field which is undergoing dramatic growth and has generated enthusiastic studies at the frontiers of organic, inorganic, supramolecular, and polymer chemistry.[1] Since the pioneering pivotal works of Vögtle[2] (*cascade synthesis*), Denkewalter,[3] Newkome[4] (*arborols*), and Tomalia[5] (*starburst dendrimers*), the synthetic research on dendritic macromole-cules has generally been centered on the creation of purely organic and more recently inorganic dendrimers with high molecular weight, and as many generations as possible. In recent years, the trends in the research on dendrimers have changed and the emphasis has been focused mainly in the modification of the properties of dendritic molecules by the introduction of functional groups on the outer dendrimer surface.[6] Since surface chemistry governs the properties of dendrimers, a rich variety of new dendritic materials explicitly designed to possess specific functions can be envisaged.

More recently, significant efforts have been also directed toward dendrimers possessing transition metals. For example, by using a multisalen dendrimer with peripheral imine functionalities Moors and Vögtle[7] prepared a trinuclear cobalt complex. In other cases, dendrimers have been constructed around transition metal-containing cores with specific properties such as luminiscence effects or electrochemical behavior that can be changed, and even modulated, with selected groups located on the dendrimer surface. In this regard, in 1992 Balzani and co-workers[8] by using the *complexes as ligands* and *complexes as metals* strategy constructed the first cascade polynuclear homo- and heterometallic (ruthenium and osmium) polypyridine complexes exhibiting the light-harvesting effect.[8c] One year later, Inoue et al.[9] reported the the first example of a photoactive metal porphyrin encapsulated in a dendritic cage. More recently, Diederich et al.[10] have evidenced that the redox chemistry, exhibited by dendritic metal porphyrins markedly differs from that of conventional non-dendritic zinc porphyrins. On the other hand, Newkome et al.[11,12] successfully achieved the controlled incorporation of transition metal centers at predetermined internal binding sites within hydrophilic-surfaced

Micellane™ dendritic frameworks, and thus prepared remarkable *Cobaltomicellanes™*[11] as well as *Rutheniomicellanes™*.[12]

In contrast, during this period much less attention was devoted to the construction of dendritic macromolecules containing organometallic entities. Nevertheless, over the last years *organometallic polymers* have emerged as an important category of new materials.[13,14] The interest in developing these materials resulted from the fact that the incorporation of transition metals into polymeric structures allows access to specialty materials with unusual and attractive characteristics including electrical, magnetic, preceramic, and catalytic properties, and nonlinear optical (NLO) effects. In many cases, however, conventional synthetic routes developed to prepare organometallic polymers unfortunately do not provide control over the molecular weight or over the position and orientation of the functional groups, and usually lead to materials either of low molecular weight, low solubility, or poorly defined structure.

On the contrary, dendritic macromolecules containing organometallic entities offer attractive advantages since they possess a precisely defined three-dimensional molecular architecture, and because of the potential to fully control their chemical constitution. Based on the well-known organotransition metal chemistry, it is clear that the incorporation of organometallic entities into dendritic structures represents a stimulating and challenging target in both organometallics and dendrimers research because it opens the way to new nanostructured organometallic macromolecules of desired nuclearity and new topologies. Likewise, the σ- or π-character of the metal–carbon bond in the organometallics, as well as the flexible coordination of transition metals and their variety of stable oxidation states, will have a significant influence on the reactivity of the dendrimer, and provide a unique opportunity for tailoring organometallic dendrimers to achieve desirable properties. Thus, by choosing the appropriate core and building blocks, dendritic macromolecules having organometallic entities in precise positions and numbers can be designed and constructed for well-defined applications (e.g., as dendritic catalysts, in multielectron redox and photocatalytic processes, as molecular sensors, and others).

This review focuses on recent published works that have given rise to the birth of a new and exciting branch of research in the field of dendrimers: *organometallic dendritic macromolecules*. The chapter intends to cover the results published as far as the early summer of 1995, and has been organized into two main sections according to the nature of the core and building blocks, as well as to the synthetic approach. As it is shown, two main strategies have been employed for the construction of such structures. First, functionalization of preformed dendrimers with reactive organometallic moieties has proven to be an excellent method for preparing new well-defined organometallic macromolecules. This is clearly illustrated by a variety of silicon-based organometallic dendrimers. In this regard, we summarize some synthetic efforts we have made toward a variety of organometallic dendrimers accessible by this approach, and in particular, the structural characterization and the redox properties exhibited by a family of ferrocenyl-containing dendrimers have been examined. Likewise, remarkable silicon-based dendrimers exhibiting catalytic activity are also described. The second approach is based on synthetic routes which start with key organometallic entities from which several expert groups have developed convergent and divergent growth strategies for the construction of very notable dendritic organometallic structures.

II. ORGANOMETALLIC DENDRIMERS CONSTRUCTED FROM ORGANOSILICON DENDRITIC CORES

A. Ferrocenyl Organosilicon Dendrimers

Ferrocenyl-containing dendrimers constructed from organosilicon dendritic frameworks represented our first target organometallic dendritic molecules. Our interest in preparing such systems stems from our investigations on silicon-containing ferrocenyl multimetallic compounds and polymers, and in particular, our efforts to understand the relationship between the structure and the redox properties of ferrocene siloxane materials.[15-17] Due to its high thermal stability and its interesting chemical and physical properties, ferrocene has become a versatile building block for the synthesis of materials with tailor-made properties.[18] Thus, ferrocene-based polymers have attracted great interest for the chemical modification of electrodes,[19] as electrode mediators,[20] and as materials for the construction of electronic devices[21] and nonlinear optical (NLO) systems.[22]

The same reasons for the interest in incorporating ferrocene units into polymers also provided motivation for the synthesis of dendritic macromolecules of well-defined size and structure containing ferrocenyl units. An important additional rationale for the construction of ferrocenyl dendrimers is provided by the fact that such macromolecules raise the possibility of combining the unique and valuable redox properties associated with the ferrocene nucleus with the highly structured macromolecular chemistry. This may provide access to materials of nanoscopic size possessing unusual symmetrical architectures, as well as specific physical and chemical properties which would be expected to differ from those of the ferrocene-based materials prepared to date.

Our initial approach to the construction of ferrocene-containing dendrimers focused on reactions that exploit the reactivity of organosilicon dendrimers functionalized at their peripheries with Si–Cl and Si–H sites toward suitable reactive ferrocenyl monomers. The divergent synthetic route to the key silicon-based dendrimers selected as frameworks starts with tetraallylsilane (**G0**) as four-directional, tetrahedral center of branching (generation 0), and is shown in Scheme 1.[23] This procedure follows approximately the valuable methodology developed by van der Made et al.,[24] Roovers et al.,[25] and Seyferth et al.[26] for the synthesis of organosilicon dendrimers. First, hydrosilylation of all allyl groups in **G0** with either chlorodimethylsilane (mono-directional branching unit) or dichloromethylsilane (two-di-

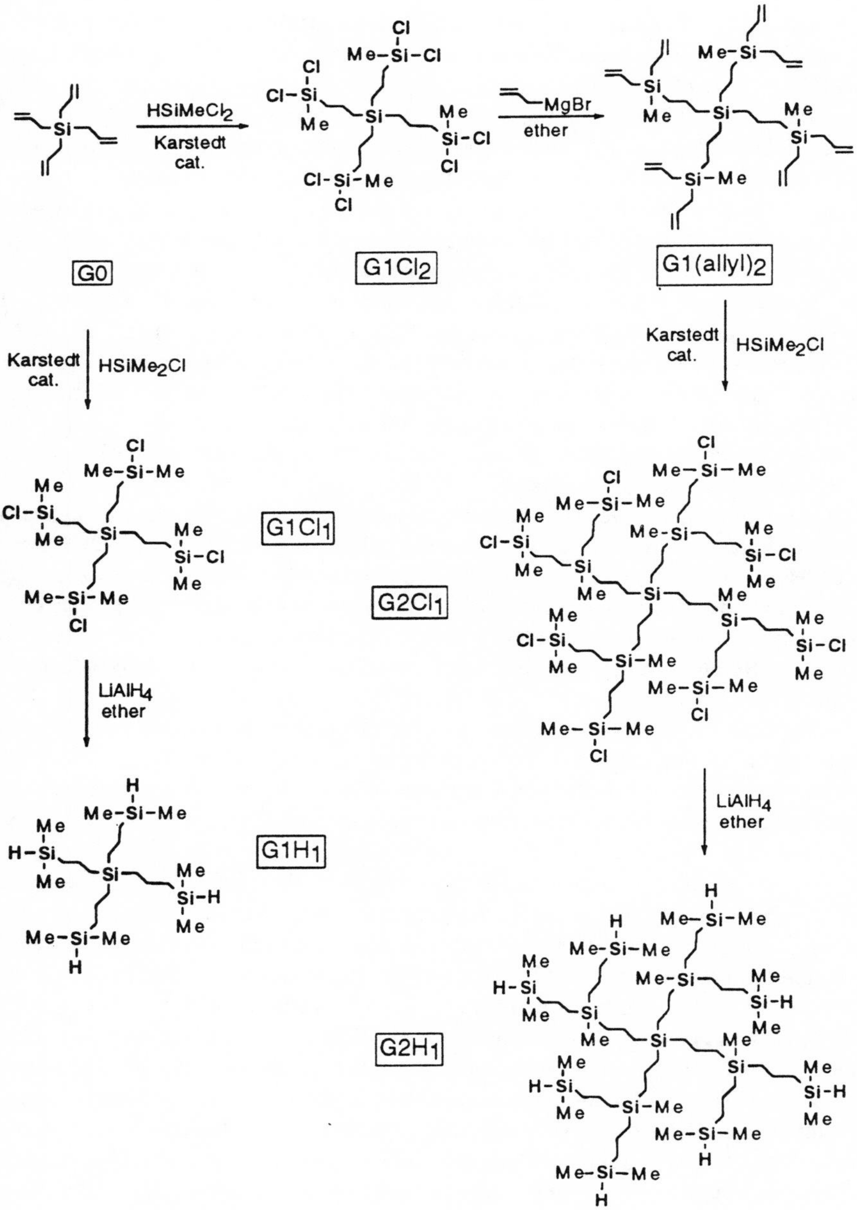

Scheme 1. Construction of organosilicon dendrimers using a tetra-directional inicia-tor core.

rectional branching unit) in the presence of Karstedt catalyst [bis(divinyltetra-methyldisiloxane)platinum(0) in xylene] afforded the silane dendrimers **G1Cl$_1$** and **G1Cl$_2$** with four SiCl and SiCl$_2$ groups, respectively. Next, growing of the branches in **G1Cl$_2$** was achieved by allylation of the SiCl groups with an excess of allylmagnesium bromide in diethyl ether at reflux, allowing the synthesis of the octakisallyl functionalized dendritic silane **G1(allyl)$_2$** which corresponds to a first allyl generation. Subsequently, all the allyl groups in **G1(allyl)$_2$** were hydrosilylated with chlorodimethylsilane in order to obtain the target dendrimer **G2Cl$_1$** with eight reactive SiCl end groups. It is interesting to note that a critical problem in carrying out hydrosilylation of allylic groups is the control of the regioselectivity of the Si–H addition in order to avoid the formation of the α-isomer and to obtain the desired β-isomer as the unique product. In our syntheses, the hydrosilylations have been carried out without a solvent in the presence of Karstedt catalyst, and according to the ^{1}H NMR spectra of the hydrosilylation products only the β-isomers were formed, which assures a regular dendrimer growth and the generation of dendritic molecules of maximum symmetry.

On the other hand, we also prepared organosilicon dendrimers possessing reactive terminal SiH groups. Reduction of the chlorosilanes **G1Cl$_1$** and **G2Cl$_1$** with LiAlH$_4$ in diethyl ether, followed by hydrolysis with dilute HCl, afforded the corresponding silicon hydrido-terminated dendrimers **G1H$_1$** and **G2H$_1$**, which bear four and eight reactive SiH moieties at their peripheries, respectively. Purification of the different **GH** as well as **Gallyl** generations was achieved by silica gel chromatography.

All the above described reactions proceed cleanly in nearly quantitative yields (80–95%). The new dendrimers display excellent solubility in most organic solvents. The structures of the novel organosilicon dendrimers were confirmed by elemental analysis, infrared spectroscopy, mass spectrometry, and ^{1}H, ^{13}C, and ^{29}Si NMR spectroscopy.

Due to the high symmetries of the molecules, the NMR spectra for these silicon-based dendrimers are relatively simple and well-defined. ^{29}Si NMR spectra are particularly useful in their characterization since they display clearly separated signals for the different types of silicon atoms in the molecules (see Figure 1). In this regard, the most interesting observation to be made is the easy assignment of the resonances corresponding to the terminal silicon atoms with different chemical environments. Thus, for **G1Cl$_1$** and **G2Cl$_1$** the outermost SiCl silicon atoms resonate at 31.11 and 31.06 ppm, respectively, far downfield of the SiAllyl silicon atoms in **G1(allyl)$_2$** which appear at 0.25 ppm, and also of the SiH silicon atoms resonances shown upfield at −14.23 and −14.07 ppm for **G1H$_1$** and **G2H$_1$**, respectively. On the other hand, the resonances corresponding to the central and middle silicon atoms, with similar chemical shifts in the expected regions of the spectrum, are assignable on the basis of the peak intensities. Thus, as the dendrimer grows the concentration of the core silicon atom becomes lower, and the intensity of the peak near 1.1 ppm for this silicon decreases from **G1** to **G2**, thus in **G2Cl$_1$** as well as in

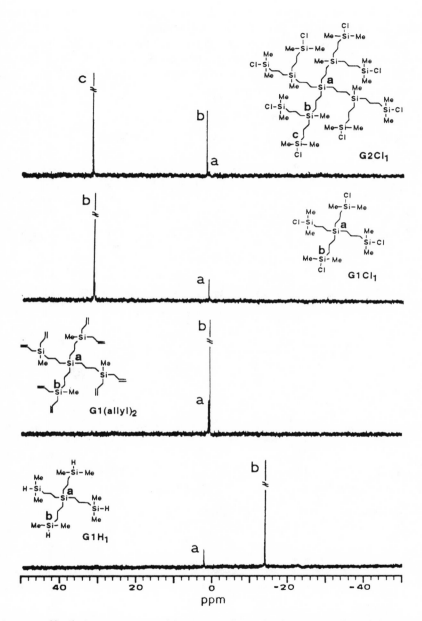

Figure 1. $^{29}Si\{^1H\}$ NMR spectra of the organosilicon dendrimers **G1Cl$_1$**, **G2Cl$_1$**, **G1H$_1$**, and **G1(allyl)$_2$**, at 59.3 MHz in CDCl$_3$.

G2H$_1$ it becomes a barely visible shoulder on the more intense peak of the next furtherout Si atom at 1.28 and 1.22 ppm, respectively.

Three different synthetic routes have been developed for the incorporation of ferrocenyl moieties onto organosilicon dendritic cores, which are shown in Scheme 2. The first method involved reactions of the tetra- and octafunctionalized **G1Cl$_1$**

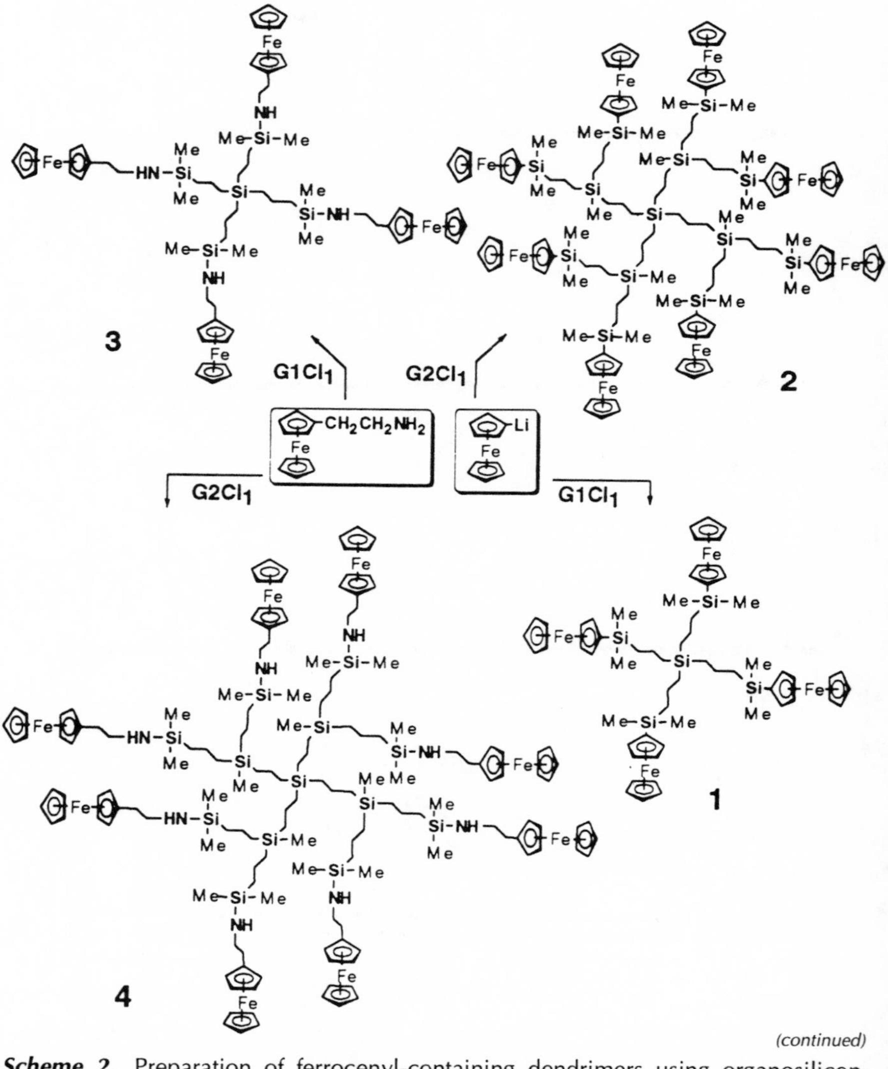

Scheme 2. Preparation of ferrocenyl-containing dendrimers using organosilicon dendritic cores.

(continued)

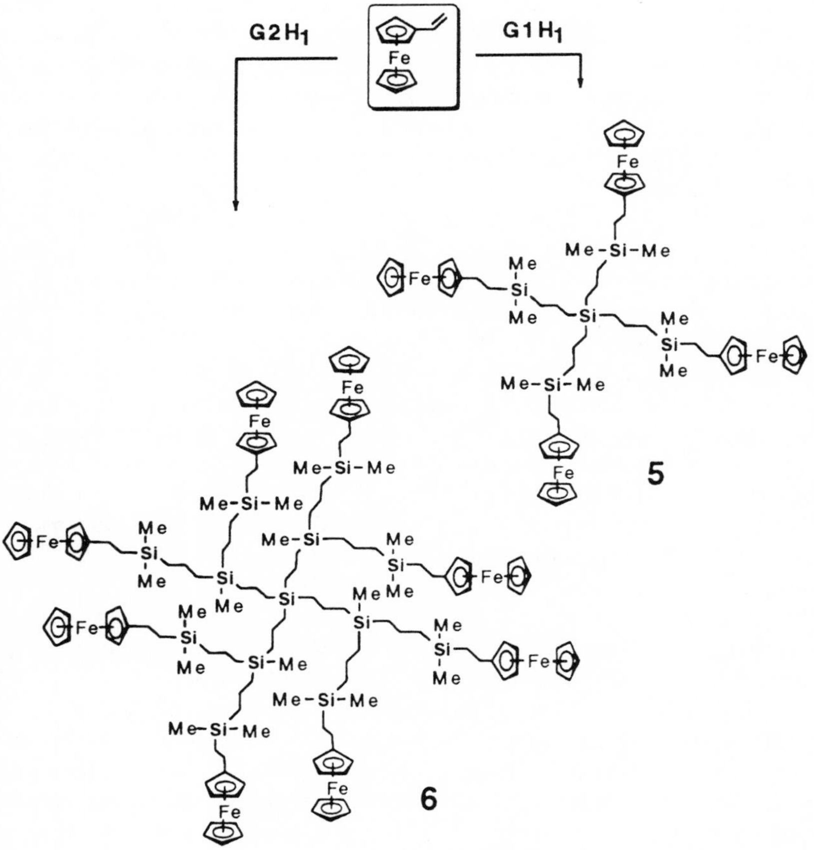

Scheme 2. Continued

and **G2Cl$_1$** with monolithioferrocene $(\eta^5\text{-}C_5H_4Li)Fe(\eta^5\text{-}C_5H_5)$ in THF at 0 °C.[23] These reactions afforded the first and second generation of dendritic macromolecules **1** and **2** which possess four and eight peripheral ferrocenyl moieties, respectively, directly bonded to the external silicon atoms of the organosilicon dendritic framework.

On the other hand, the high reactivity of the Si–Cl bonds toward the amine groups also allowed facile organometallic functionalization of the surface of dendrimers **G1Cl$_1$** and **G2Cl$_1$**. The key monomer in this synthesis was (β-aminoethyl)ferrocene $(\eta^5\text{-}C_5H_4CH_2CH_2NH_2)Fe(\eta^5\text{-}C_5H_5)$, which was selected because the amino reactive group is two methylene units removed from the ferrocene nucleus. This fact is of critical importance because it minimizes steric and electronic effects due to the organometallic moiety, and likewise, the instability found in α-functional ferrocene derivatives, due to the α-ferrocenyl carbonium ion stability is removed.[27] Treatment of **G1Cl₁** and **G2Cl₁** with the appropriate mole ratios of (β-aminoethyl)ferrocene in toluene at reflux temperature and in the presence of triethylamine as the acid acceptor yielded the desired tetra- and octanuclear macromolecules **3** and **4**, which remarkably possess N–H linked ethylferrocenyl moieties attached to the surface of the dendritic structure.

We also explored a different synthetic route to build up peripherally functionalized polyferrocenyl dendritic systems involving the use of hydrido-terminated organosilicon dendrimers. This strategy is based on our earliest successful synthetic efforts focused toward the incorporation of ferrocenyl moieties into polyfunctional cyclic and polyhedral siloxane frameworks via hydrosilylation reactions of vinylferrocene $(\eta^5\text{-}C_5H_4CH{=}CH_2)Fe(\eta^5\text{-}C_5H_5)$ with the SiH-containing 1,3,5,7-tetramethylcyclotetrasiloxane and octakis(hydrodimethylsiloxy)octasilsesquioxane.[15,17] In view of these results, the hydrosilylations of both **G1H₁** and **G2H₁** with the appropriate equivalents of vinylferrocene in toluene at 60 °C and in the presence of catalytic amounts of Karstedt catalyst were investigated. These reactions cleanly afforded the novel ferrocene dendrimers **5** and **6** (Scheme 2) in which the external silicon atoms of the dendritic core units are attached to ferrocenyl moieties through a two-methylene flexible spacer.

The three families of ferrocene dendritic macromolecules described above, in which the ferrocenyl units are at the end of long flexible silicon-containing chains, are air-stable red-orange thick materials. They present several significant advantages including the relative ease of preparation, and the remarkably high solubility in typical organic solvents such as dichloromethane, THF, and even in nonpolar hydrocarbons such as hexanes. Indeed, the solubility of these dendrimers does not seem to diminish with increasing generations. The purification procedures depended on the reaction procedure, and in general involved hydrolytic workup, sublimation in vacuum of unreacted starting ferrocene or some reaction by-products, and repeated column chromatographies on silanized silica.

It is noteworthy to mention that the chemistry necessary to perform the syntheses described above appeared to be well in hand and suitable for incorporation of organometallic units into dendrimeric structures. Likewise, the synthetic routes we have employed are of potentially broad applicability for designing novel organometallic-containing dendrimers because they permit structural variation in the starting organosilicon frameworks as well as variations of the organometallic functionality. Thus, by changing the multiplicity of the initiator core, the degree of branching, and the length of the branches it is possible to construct compounds in which different organometallic centers occupy predetermined positions in the macromolecular array.

The structures of the novel ferrocenyl dendrimers **1–6** were straightforwardly established by a variety of spectroscopic and analytical techniques including ^{1}H, ^{13}C, and ^{29}Si NMR and IR spectroscopies, fast atom bombardment mass spectrometry (FABMS), molecular weight determination (VPO), as well as elemental analysis. The high symmetry of these macromolecules has made NMR spectroscopy a useful technique for their characterization since only uniform growth and pure products give simple and well-defined NMR spectra.

For example, in the ^{1}H NMR spectra, confirming evidence for the complete functionalization of the four and eight reactive sites in the organosilicon dendritic cores with ferrocenyl moieties is provided by the ratio of integration of the ferrocenyl and methylene protons, which agrees with the calculated values. The ^{29}Si NMR spectra are also of interest as they exhibit the signals expected for the different types of silicon present in the dendrimers. For instance, in the spectrum of dendrimer **6** shown in Figure 2, although the chemical environments around the three different silicon atoms are very similar, the corresponding peaks appear clearly separated and can be easily assigned on the basis of the peak intensities.

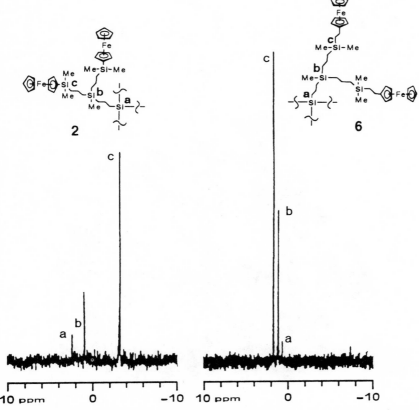

Figure 2. ^{29}Si{^{1}H} NMR spectra of the second generation octaferrocenyl dendrimers **2** and **6**, at 59.3 MHz in CDCl$_3$.

Likewise, three resonances can be also observed in the ^{29}Si NMR spectrum of **2**, but in this case the resonance of the outermost silicon atoms directly bonded to the ferrocenyl moiety is shifted upfield and occurs at −3.0 ppm.

Attempts to isolate the ferrocenyl dendrimers described above in a crystalline form suitable for X-ray structural determination have so far been unsuccessful. For this reason, we have used computer-generated molecular models in order to gain further information about the structural features of these materials. Figure 3 illustrates an energy-minimized structure determined from CAChe™ molecular mechanic calculations of the ferrocenyl dendrimer **2**. From these studies, we have measured approximate diameters of 2 nm for the first-generation dendrimers **1**, **3**, and **5**, and 3 nm for the second-generation dendrimers **2**, **4**, and **6**.

On the other hand, we have also isolated and characterized the oxidized forms of some of the synthesized polyferrocene dendrimeric systems.[28] Chemical oxidation of the ferrocenyl units in **1**, **2**, **5**, and **6** has been achieved by reaction with $NOPF_6$ in CH_2Cl_2 solution, and affords the blue tetra- and octanuclear dendritic cations $[\mathbf{1^{4+}}][\mathbf{PF_6^-}]_4$, $[\mathbf{2^{8+}}][\mathbf{PF_6^-}]_8$, $[\mathbf{5^{4+}}][\mathbf{PF_6^-}]_4$, and $[\mathbf{6^{8+}}][\mathbf{PF_6^-}]_8$. The UV−visible spectra of CH_2Cl_2 solutions of these cations (see Figure 4A) consist of two strong bands at about 250 and 290 nm which correspond to allowed charge-transfer transitions, and a weak band at 625 nm assigned to a ligand to metal charge-transfer transition

Figure 3. Structure of dendrimer **2** from molecular modeling.

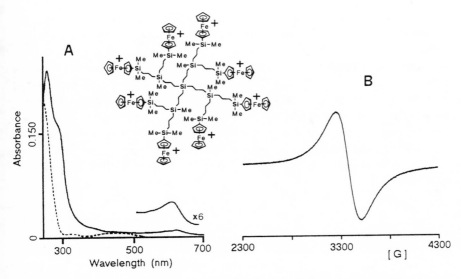

Figure 4. (A) UV-Visible spectra in CH_2Cl_2 solution of the octanuclear dendrimers $[2^{8+}][PF_6^-]_8$ (——), and **2** (- - -). (B) X-band EPR spectrum for the cationic dendrimer $[2^{8+}][PF_6^-]_8$, recorded in CH_2Cl_2 glass at 90 K.

associated with the ferrocenium cations,[29a,b] which is absent in those of the neutral dendrimers.

Likewise, we have further characterized the dendritic cations by electron paramagnetic resonance spectroscopy (EPR). Of particular interest is the fact that the EPR spectra of CH_2Cl_2 frozen solutions (liquid nitrogen, 90 K) of the PF_6^- salts of all the ferrocenium dendrimers (see for example Figure 4B) exhibit one isotropic broad signal ($\Delta H = 255$ G for **1**, $\Delta H = 320$ G for **2**, $\Delta H = 275$ G for **5**, and $\Delta H = 345$ G for **6**) with *g* values of 2.00, 2.03, 1.99, and 1.99, respectively, indicating that these polyferrocenium species have the unpaired electron essentially localized on the silicon-substituted cyclopentadienyl fragment.[29c–e] The EPR spectral isotropic feature observed at 90 K is remarkable, and is consistent with ferrocenium moieties in highly symmetric environments.

Electrochemical Properties

The incorporation of redox-active organometallic units within or on the periphery of dendritic structures is an especially challenging target because such molecules are good candidates to play a key role as multielectron transfer mediators in electrocatalytic processes of biological and industrial importance. In particular, the organometallic ferrocene moiety is an attractive redox center to integrate into dendritic structures, not only because it is electrochemically well behaved in most common solvents undergoing a reversible one-electron oxidation, but in addition such electron removal commonly does not involve fragmentation of the original molecular framework.[30]

The redox behavior of the peripherally functionalized ferrocenyl dendritic macromolecules 1–6 has been studied in CH_2Cl_2 with tetrabutylammonium hexafluorophosphate (TBAH) as the supporting electrolyte. The cyclic voltammograms of the tetranuclear dendrimers 1, 3, and 5, as well as those of the octanuclear 2, 4, and 6 show a single reversible oxidation process (see Figure 5), at potentials that are a function of the electron-donating ability of the different substituents attached to the cyclopentadienyl rings. The results of cyclic voltammetry (CV), differential pulse voltammetry (DPV), and controlled potential electrolysis clearly show that in the tetra- and octanuclear dendrimers the observed reversible oxidation waves represent a simultaneous multielectron transfer of four and eight electrons, respectively, as expected for independent reversible one-electron process, at the same potential, of the four (in 1, 3, and 5) or eight (in 2, 4, and 6) ferrocenyl moieties. We particularly note that for dendrimers 1, 2, 5, and 6 careful coulometry measurements not only result in the removal of the expected number of electrons for all the peripheral ferrocenyl groups (4.0 or 8.0 electron/molecule), but in addition reelectrolysis quantitatively regenerated the starting neutral polynuclear compounds. This result contrasts interestingly with that observed in organic dendrimers functionalized with redox-active tetrathiafulvalene (TTF) units,[6d] in which oxidation of only half of the non-interacting TTF units was observed using a range of electrochemical techniques. On the other hand, the diffusion coefficients of the ferrocenyl-containing

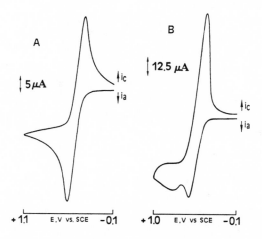

Figure 5. Cyclic voltammograms of dendrimers (**A**) **6** and (**B**) **4**, in solution of CH_2Cl_2/0.1 M TBAH, at a scan rate of 50 mV/s.

dendrimers were calculated using the Randles–Sevcik equation,[31] resulting, for example, in $D_o = 1.34 \times 10^{-6}$ cm^2 s^{-1} for **1**, and $D_o = 8.58 \times 10^{-7}$ cm^2 s^{-1} for **2**.

Of particular relevance is that the solution redox behavior of dendrimers **3** and **4**, which possess electroactive ethylferrocenyl moieties bonded to the dendritic framework through N–H linkages, is different from that observed in the other ferrocenyl dendrimers (Figure 5B). First, an additional minor oxidation wave is observed at a more anodic redox potential than that of the ferrocene/ferrocenium couple, which corresponds to the oxidation of the amino group. Furthermore, changes in solubility with the change in the oxidation state in the ferrocenyl units were observed for **3** and **4** in CH$_2$Cl$_2$/TBAH solution. In fact, a sharp reduction peak is observed which is characteristic of redox couples in which the oxidized form is insoluble and the reduced form is soluble—that is, a stripping peak. In addition, upon continuous scanning there is an increase in the peak current with each successive scan, indicating that formation of an electroactive dendrimer film occurs on the electrode surface.[32]

Without doubt, the most noteworthy aspect of the redox behavior of the synthesized organometallic dendritic macromolecules **1–6**, having a predetermined number of noninteracting ferrocenyl redox centers, is their ability to modify electrode surfaces. In this way, for the first time, electrode surfaces have been successfully modified with films of dendrimers containing reversible four- and eight-electron redox systems, resulting in detectable electroactive materials persistently attached to the electrode surfaces.[28,32]

The ferrocenyl dendrimers were electrodeposited in their oxidized forms onto the electrode surfaces (platinum, glassy-carbon, and gold) either by controlled potential electrolysis or by repeated cycling between the appropriate anodic and cathodic potential limits; therefore the amount of electroactive material electrodeposited can be controlled with the electrolysis time or the number of scans. The electrochemical behavior of films of the polyferrocenyl dendrimers was studied by cyclic voltammetry in fresh CH$_2$Cl$_2$ and CH$_3$CN solutions containing only supporting electrolyte.

The voltammetric response of an electrodeposited film of **2** in CH$_2$Cl$_2$ with 0.1 M TBAH is shown in Figure 6 as a representative example. A well-defined, symmetrical oxidation–reduction wave is observed, which is characteristic of surface-immobilized reversible redox couples, with the expected linear relationship of peak current with potential sweep rate v.[33] A formal potential value of $E^o_{surf} = +0.42$ V vs. SCE was found for the surface-confined octaferrocenyl dendrimer **2** that is nearly identical to the formal potential of the macromolecule in solution ($E^o = +0.43$ V vs. SCE in CH$_2$Cl$_2$/TBAH). The electroactive dendrimer film behaved almost ideally with rapid charge transfer kinetics. These voltammetric features unequivocally indicate the surface-confined nature of the electroactive ferrocenyl moieties in the dendrimer. Surface coverage of electroactive ferrocenyl sites in the film Γ(mol cm^{-2}) was determined from the integrated charge of the cyclic voltammetric wave. The obtained results correspond to about a close-packed monolayer of

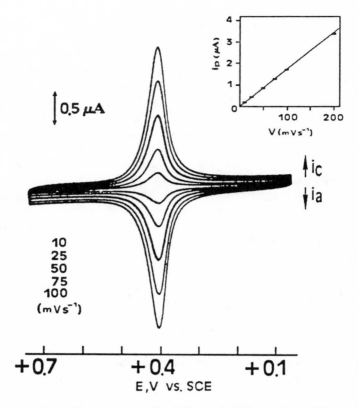

Figure 6. Cyclic voltammograms of a platinum disk-electrode modified with a film of the octanuclear dendrimer **2**, measured in 0.1 M Bu₄NPF₆/CH₂Cl₂. The surface coverage of electroactive ferrocenyl sites in the film is determined to be $\Gamma = 2.01 \times 10^{-10}$ mol cm^{-2}. *Inset*: scan rate dependence of the anodic peak current.

ferrocene units in the surface. One of the most remarkable features of Pt electrodes modified with films of these ferrocenyl dendritic macromolecules is that they are extremely stable and reproducible. Indeed, cyclic voltammetric scans can be carried out in either organic or aqueous electrolyte solutions hundreds of times with no loss of electroactivity. The electroactivity of the dendrimer-modified electrodes was retained after storage in air several weeks after preparation.

The microstructure of films of the ferrocene dendrimers electrochemically deposited on platinum wire working electrodes was examined by scanning electron microscopy (SEM). The SEM micrograph in Figure 7 corresponding to a film of the octanuclear dendrimer **2** shows a sheet-like compact morphology and exhibits small agglutinations and some porosity.

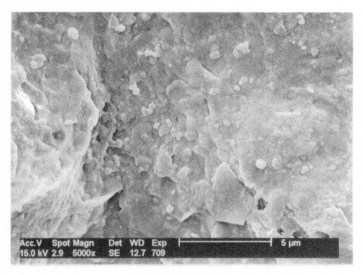

Figure 7. Scanning electron micrograph of a film of **2** electrochemically deposited on a platinum wire electrode (0.25 mm of diameter).

Ferrocenyl dendrimers also afford electroactive films on indium tin oxide (ITO) electrodes in the same manner as described above. UV–visible spectroelectro-chemical measurements of this modified electrodes on oxidation show changes characteristic for the formation of ferrocenium cations. Thus, Figure 8 shows the UV–visible absorption spectrum of a film of **2** electrodeposited on a transparent ITO electrode, which exhibits a strong band at 260 nm and a weak absorption band centered at 600 nm, which agree with those observed for the cationic dendrimer $[2^{8+}][PF_6^-]_8$ in solution described above.

In conclusion, we have demonstrated the feasibility of modifying electrode surfaces with organometallic dendrimers bearing a predetermined number of equivalent redox centers.

Application of Ferrocenyl-Containing Dendrimers in the Electrochemical Recognition of Anions and as Electron Transfer Mediators in Amperometric Biosensors

The molecular recognition of anionic guest species by positively charged or neutral receptors is a relatively new area of research of growing interest[34] in view of the key roles that these anions play in biochemical and chemical processes. For this reason, as part of the electrochemical studies, we decided to examine the use of the redox-active ferrocenyl dendrimers **3** and **4** that contain multiple N–H linkages capable of participating in H-bonding, as well as characteristic internal cavities,

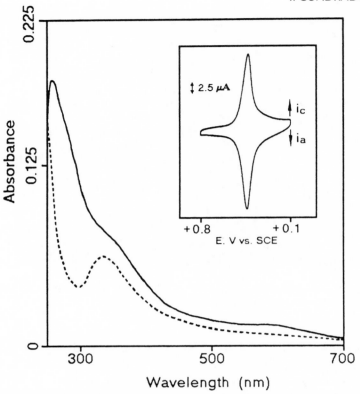

Figure 8. UV-visible spectra of a film of **2** electrodeposited on an ITO electrode, recorded after the film was oxidized at + 0.70 V (———) and subsequently reduced at 0.0 V (- - -). *Inset*: cyclic voltammogram of an ITO electrode modified with a film of **2** measured in Bu_4NPF_6/CH_2Cl_2.

both potentially suitable binding sites as redox-responsive systems for electro-chemical recognition of anions.[32]

Cyclic voltammograms of **3** and **4** in CH_2Cl_2/TBAH solution, recorded after successive addition of stoichiometric amounts of the tetrabutylammonium salts of Cl^-, Br^-, HSO_4^-, and $H_2PO_4^-$ revealed anion guest-induced variation of the potential and current peak of the ferrocenyl oxidation; and the results are shown in Figure 9 for the octanuclear dendrimer **4**. It is noteworthy that the largest magnitude of cathodic shift was observed in both compounds with the $H_2PO_4^-$ anion (about 300 mV). These results suggest that dendrimers **3** and **4** are electrochemically sensitive to the presence of such anions, probably due to the amine group participation in H-bonding.[34a] Interestingly, in the presence of the $H_2PO_4^-$ anion the shape of the oxidation wave changes from a reversible redox process to an EC mechanism as the concentration of the anionic guest increases. Likewise, when an equimolar

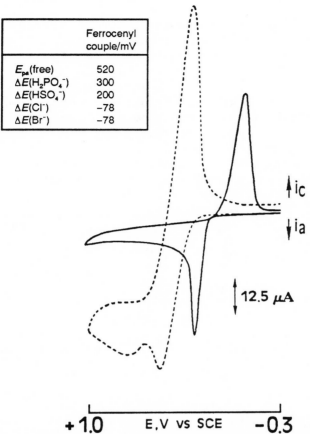

	Ferrocenyl couple/mV
E_{pa}(free)	520
$\Delta E(H_2PO_4^-)$	300
$\Delta E(HSO_4^-)$	200
$\Delta E(Cl^-)$	−78
$\Delta E(Br^-)$	−78

Figure 9. Cyclic voltammograms of **4** in Bu_4NPF_6/CH_2Cl_2, in the absence (- - -), and in the presence (——) of $H_2PO_4^-$ anion. *Inset*: electrochemical data for dendrimer **4**. Shifts of the anodic current peak potential of the ferrocene/ferrocenium couple produced by the presence of anions (2 equiv.) added as their tetrabutylammonium salts ($\Delta E = [E_{pa}$(free) $- E_{pa}$(anion)]).

mixture of HSO_4^- and $H_2PO_4^-$ anions is added to CH_2Cl_2/TBAH solutions of the octanuclear dendrimer **4**, the cathodic shift measured for the anodic peak potential is very similar to that induced by the $H_2PO_4^-$ anion alone. Furthermore, these preliminary studies clearly show that the voltammetric response of electrodes derivatized with electroactive films of **4** is sensitive to the presence and concentration of the $H_2PO_4^-$ anion. These first electrochemical results clearly suggest that the polyferrocenyl NH-containing dendritic molecules **3** and **4** can represent novel type of acyclic receptor systems for the electrochemical recognition of anions both in solution and at the electrode–solution interface; further work on this interesting subject is currently in progress.

Over the last few years, new electrodes using enzymes as sensing elements have been developed. Amperometric enzyme electrodes with electroactive species acting as mediators replacing the natural electron acceptor, dissolved oxygen, have been widely investigated.[35] Ferrocene and its derivatives have been shown to be used successfully as mediators in various enzyme electrodes.[20] Potentially, the synthesized ferrocene dendritic macromolecules, in which the ferrocenyl units are at the end of long flexible silicon-containing branches, can serve to wire electrically the enzyme, facilitating a flow of electrons from the enzyme to the electrode. For this reason, in order to test the ability of ferrocenyl dendrimers to act as electron mediating species, a study of the efficiency of dendrimers/glucose–oxidase/carbon paste electrodes was undertaken.[36] Some of the most significant results of this study are illustrated in Figure 10 for dendrimers **2**, **5**, and **6**.

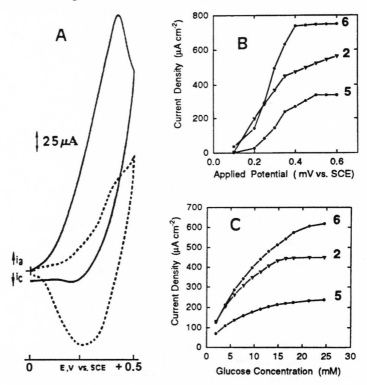

Figure 10. (**A**) Cyclic voltammograms for the dendrimer **6**/glucose oxidase/carbon paste electrode, recorded at 1 mV/s in pH 7.0 phosphate buffer (with 0.1 M KCl) solution, with no glucose present (- - - -), and in the presence of 0.1 M glucose (———). (**B**) Steady-state polarization curves of carbon paste electrodes in the presence of 33.3 µM glucose. (**C**) Variation of the steady-state current of carbon paste electrodes with glucose concentration (at +350 mV vs. SCE).

Cyclic voltammograms of these carbon paste electrodes show that the addition of glucose leads to the enhancement of the oxidation current, while cathodic current is not observed (see e.g. Figure 10A). This fact is indicative of enzyme-dependent catalytic reduction of the ferrocenium cations. In addition, the electrodes are clearly sensitive to small changes in glucose concentration and display a good response over long periods of time. As shown in Figures 10B and 10C, for equimolar amounts of the ferrocene moieties, the octanuclear dendrimers **2** and **6**, possessing the longer organosilicon branches, have proved to mediate electron transfer more efficiently than the relay systems based on the tetranuclear **5**. On the other hand, it is clear that dendrimer **6**, in which the ferrocenyl units are attached to the dendritic framework through a two-methylene flexible spacer, is more effective at mediating electron transfer between reduced glucose oxidase and the carbon paste electrode. These results clearly suggest that the flexibility of the dendritic mediator is an important factor in the ability to facilitate the interaction between the mediating species and the flavin adenine dinucleotide (FAD) redox centers of glucose oxidase.

Hyperbranched Ferrocenyl Silicon-Based Polymers

The ferrocene dendritic macromolecules described above provide excellent models for the structural and electrochemical properties of a new class of highly branched ferrocene polymers. Bearing this in mind, and in order to complete this ferrocenyl dendrimeric series, we have extended our preparations to more complex polymeric structures by assembling the synthesized organosilicon dendrimers through suitable difunctional organometallic fragments. This has involved the preparation of the three-dimensional organometallic networks **7** and **8** shown in Figure 11. Their synthesis was achieved by a salt-elimination reaction of dilithioferrocene · TMDA (TMDA = tetramethylethylenediamine) with the chlorosilane **G1Cl₁** in hexane, and Pt-catalyzed hydrosilylation of 1,1'-divinylferrocene with the silicon hydride **G1H₁** in toluene, respectively. Interestingly, these highly branched polymeric materials show a greater solubility than that observed in related linear silicon-containing ferrocenyl polymers,[16] and they are even readily soluble in nonpolar hydrocarbons, such as *n*-hexane. The novel dendritic materials **7** and **8** have a three-dimensional branching structure, which is in concept related to those shown by two other organometallic network polymeric structures **9** and **10** derived from a cyclotetrasiloxane and an octasilsesquioxane, respectively, that we have reported recently.[15,17]

The thermal behavior of the hyperbranched polymer **7**, as well as that of the related tetra- and octanuclear dendrimers **1** and **2**, was examined by thermal gravimetric analysis (TGA) and derivative thermogravimetry (DTG). The samples were heated at a ramp rate of 10 °C/min under nitrogen atmosphere in the temperature range 25–800 °C. The major decompositions occur between 340 and 540 °C for the tetranuclear **1**, and between 370 and 520 °C for the octanuclear **2**, whereas for polymer **7** a faster rate of weight loss is observed at a higher (480 and 560 °C) temperature range. The thermal stability of the network dendrimeric material **7** is

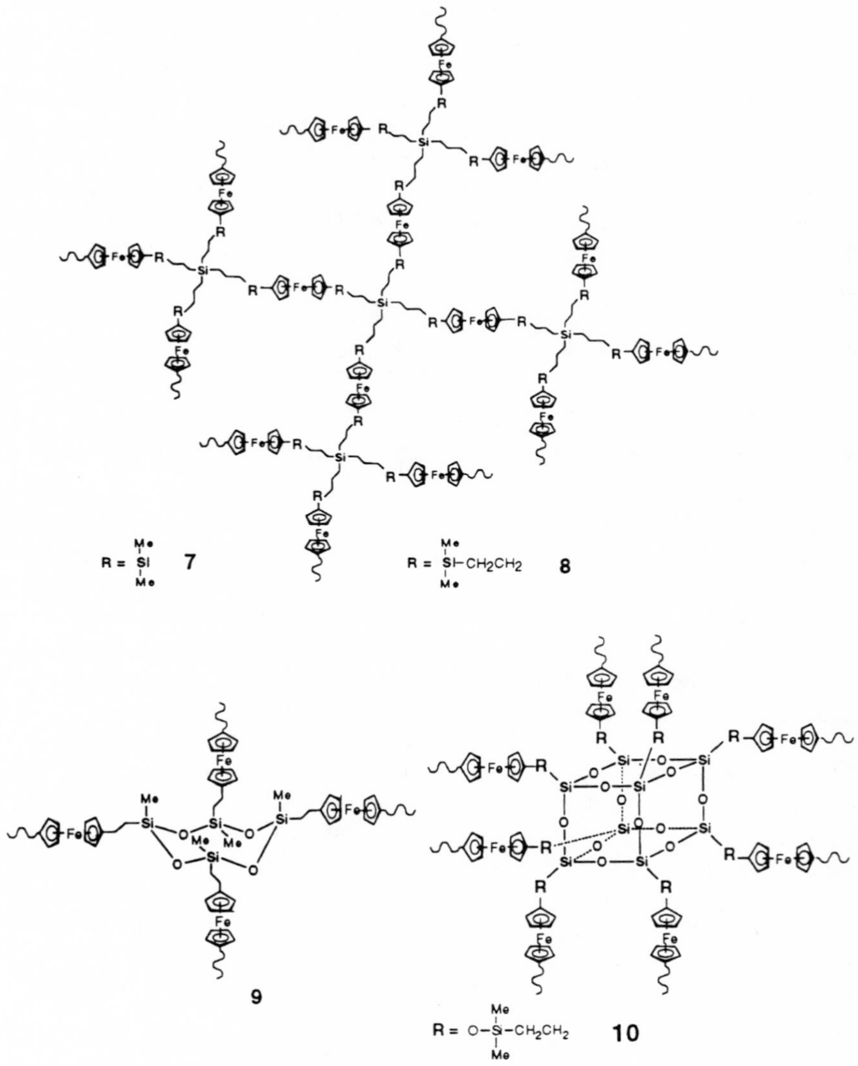

Figure 11. Schematic representation of three-dimensional ferrocene networks built from dendritic (**7** and **8**), cyclic (**9**), and polyhedral (**10**) silicon-containing building blocks.

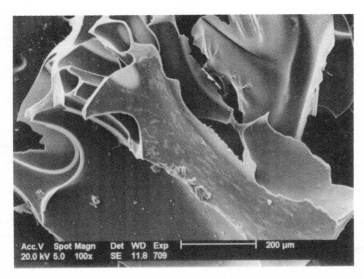

Figure 12. Micrograph of the ceramic residue obtained on pyrolysis of **7**.

comparable to that observed in related ferrocenyl polymers having siloxane frameworks such as **9** and **10**.[15,17]

One of the driving forces behind recent interest in organometallic polymers is their potential use as pyrolytic precursors of transition metal-containing ceramics which exhibit a wide range of interesting electrical, magnetic and optical properties.[14,18] In concept, the synthesized silicon-based ferrocenyl hyperbranched polymers, as well as the polynuclear organometallic dendrimers, could be pyrolytic precursors to iron silicon carbide ceramics. The black ceramic residues, obtained when **2**, **7**, and **8** are heated at 750 °C under a nitrogen stream, show magnetic properties and were characterized as iron silicon carbide materials by SEM and energy-dispersive X-ray (EDX) analysis. A representative scanning electron micrograph of the material resulting of polymer **7** is shown in Figure 12.

On the basis of the results described here, extension of these syntheses to higher generations, as well as work aimed at more detailed studies with respect to the electrochemical properties and valuable applications of these silicon-based ferrocenyl dendrimers, clearly represent interesting areas of future research.

B. Chromium-, Cobalt-, and Iron Carbonyl-Containing Dendrimers

{η⁶-*Organosilyl Arene}Chromium Tricarbonyl Dendrimers*

In order to search new synthetic strategies to build up novel families of organometallic-containing dendritic macromolecules, we explored pathways that would allow the efficient π-coordination of transition metals to terminal arene ligands on organosilicon dendrimers. Particularly, we focused on polynuclear chromium carbonyl derivatives in which $Cr(CO)_3$ moieties are π-bound to peripheral arene rings directly attached to organosilicon dendritic core units. Our interest in these systems derives from our earlier investigations of a series of chromium tricarbonyl mono-, di-, and polymetallic compounds containing reactive organosilane groups on the arene rings.[37] {η^6-Arene}$Cr(CO)_3$ complexes have been studied in a variety of applications ranging from their use as (1) stoichiometric asymmetric reagents, (2) models for ligand substitution reactions, (3) efficient catalysts in several processes, and (4) precursors of hard coatings prepared by chemical vapor deposition.[38]

The key starting organosilicon dendrimers functionalized at the periphery with phenyl rings required for this study were **G1Ph$_1$** and **G2Ph$_1$** (Scheme 3). They were prepared without difficulty by hydrosilylation of **G0** and **G1(allyl)$_2$**, respectively, with phenyldimethylsilane in the presence of Karstedt catalyst using toluene as solvent, and their purification was achieved by silica gel chromatography. The yield of **G2Ph$_1$** (50%) was significantly lower than that of the tetraphenyl **G1Ph$_1$** (93%),

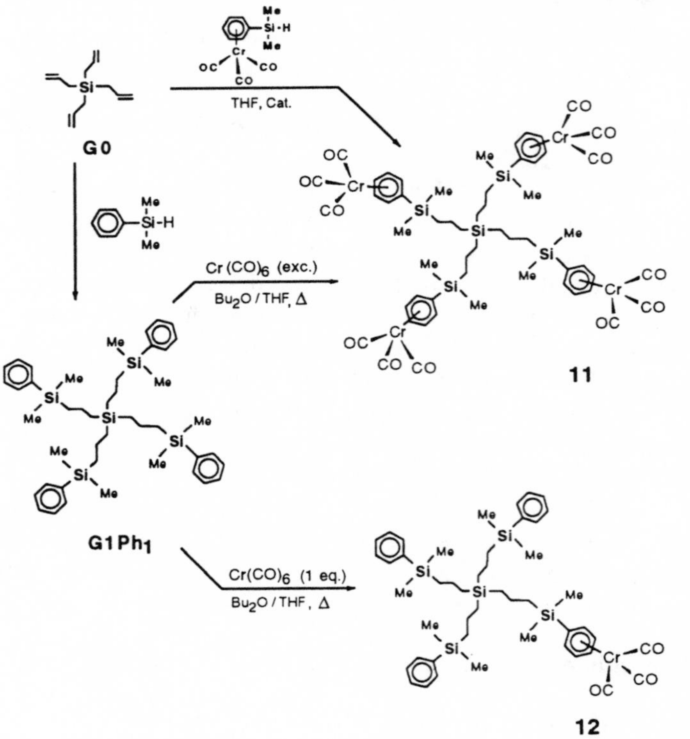

Scheme 3. Synthesis of phenyl-terminated silicon-based dendrimers and functionalization of the branches with $Cr(CO)_3$ units.

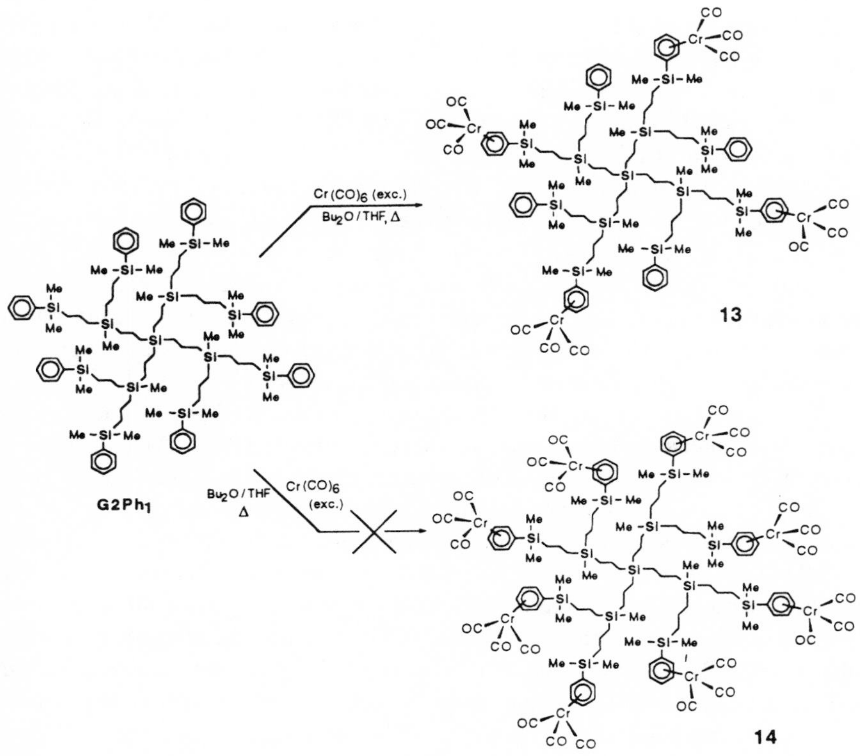

Scheme 3. Continued

probably as a consequence of the decreased reactivity of the arms in dendrimer **G1(allyl)₂** due to the enhanced steric congestion of the eight allyl substituents compared to the more sterically accessible tetrafunctionalized **G0**. In addition, some losses of product during the purification of this larger dendrimer are possible.

Our first synthetic approach for incorporating $Cr(CO)_3$ moieties into organosilicon dendritic polyfunctional cores involved thermal replacement of CO from chromium hexacarbonyl by the phenyl-terminated dendrimers **G1Ph₁** and **G2Ph₁** (Scheme 3).[39] Thus, treatment of **G1Ph₁** with an excess of $Cr(CO)_6$ in the donor solvent medium of dibutyl ether/tetrahydrofuran (9/1) at 140 °C affords the tetranuclear **11** in 70% yield which, after recrystallization from dichloromethane/hexane, was isolated as moderately air-stable, yellow crystals. Similarly, by using the appropriate equivalents of $Cr(CO)_6$ and controlling the reaction times, the mononuclear tricarbonyl macromolecule **12** was also prepared. In addition, we devised an alternate route to **11**, which involves the previous synthesis of the SiH-functionalized reactive monomer $\{\eta^6\text{-}C_6H_5Si(Me)_2H\}Cr(CO)_3$,[37] suitable for its subsequent attachment to the four-directional core **G0** via hydrosilylation.

We have also tried to extend the former thermal procedure to the next higher generation but, unfortunately, we have not succeeded in the functionalization of the eight phenyl-terminated dendrimer arms of **G2Ph$_1$** with Cr(CO)$_3$ units. Instead, from the reaction of an excess of Cr(CO)$_6$ with **G2Ph$_1$**, the tetranuclear derivative **13** was isolated as the major reaction product. The failure to obtain the desired octachromium macromolecule **14** was presumably caused by the more forcing reaction conditions (considerably longer reaction periods with elevated temperatures) required for the incorporation of high loadings of Cr(CO)$_3$ units from Cr(CO)$_6$, which produce noticeable decomposition. In order to obtain the target macromolecule with eight Cr(CO)$_3$ moieties, work is in progress aimed to overcome these problems by using either highly reactive derivatives such as (MeCN)$_3$Cr(CO)$_3$ and milder reaction conditions, or photochemical procedures.

As with the ferrocenyl-containing silicon dendrimers, the high symmetry of the tetranuclear molecule **11** makes structural verification by NMR spectroscopy very simple. For example, in its ^{1}H NMR spectrum (Figure 13C) the complete η^6-coordination of the Cr(CO)$_3$ moieties to the four phenyl rings in the dendritic silicon core was evidenced by the total absence of resonances in the range 7.37–7.53 ppm, in which the arene ligand resonances of the noncoordinated organosilicon dendrimer are observed (see Figure 13A). As a consequence of the withdrawal of the *p*-electrons from the arene rings by the Cr(CO)$_3$ moieties, the aromatic protons on the complexed dendrimer resonate at a significantly higher field, in the region from 5.6 to 5.1 ppm. In addition, the integration of resonances further confirms the tetrafunctionalization. Likewise, the ^{1}H NMR spectrum of dendrimer **12** (Figure 13B) has been particularly useful to verify the loading of Cr(CO)$_3$ units on the dendrimer. Thus, the integration ratio of the protons in the complexed ring and those of the uncomplexed aryl moieties are in agreement with the proposed monometallic structure. Further evidence for the dendritic structures was also provided by ^{13}C and ^{29}Si NMR spectroscopy, FAB mass spectrometry, and elemental analysis.

The electrochemical behavior of the mono- and tetranuclear dendrimers **11** and **12** was studied by cyclic voltammetry in CH$_2$Cl$_2$/0.1M TBAH. Both dendrimers exhibit a chemically reversible oxidation wave ($E_{1/2}$ = +0.78 and +0.80 mV vs. SCE, respectively) corresponding to the oxidation of the peripheral (η^6-C$_6$H$_5$)Cr(CO)$_3$ units. Evaluation of the number of electrons transferred in the processes was effected from the intensity of the cyclic voltammetric waves. The results indicate that, for the tetranuclear **11**, the oxidation wave corresponds to the transfer of four electrons at the same potential, affording the cationic dendritic species [**11^{4+}**] which is stable in the cyclic voltammetric time scale. Likewise, the four tricarbonylchromium moieties are essentially noninteracting redox centers. Furthermore, in the presence of tertiary phosphite nucleophiles, such as P(OBu)$_3$, the cationic species [**11^{4+}**] undergoes rapid CO substitution in an electrochemically induced process.

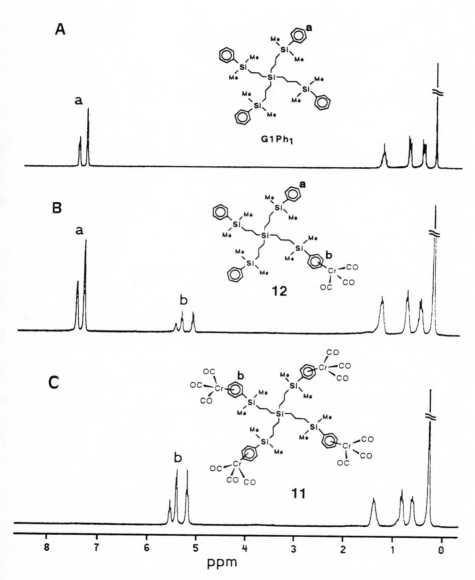

Figure 13. ¹H NMR spectra of the dendrimers **G1Ph₁**, **12** and **11**, at 360 MHz in CDCl₃.

Iron Carbonyl- and Cobalt Carbonyl-Containing Dendrimers

Within organometallic chemistry, cyclopentadiene is perhaps the most important and widely used ligand.[40] More than 80% of all known organometallic complexes of the transition metals contain the cyclopentadienyl moiety or a derivative thereof. Indeed, cyclopentadienyl derivatives, which are π- or σ-bonded to the central atom, are known for all transition metals. For this reason we thought that a stimulating challenge in the synthetic research on organometallic dendrimers would be the incorporation of the cyclopentadienyl ligand into well-defined dendritic structures because it offers an enhanced variability in the designed generation of novel families of highly structured organometallic macromolecules.

Thus, the reaction of the sodium cyclopentadienide anion with the four-directional Si–Cl dendrimer **G1Cl$_1$** in THF solution was successfully performed affording the desired cyclopentadienyl-functionalized organosilicon dendrimer **15**, shown in Scheme 4.[41] ^{1}H NMR spectrum of this polyfunctional molecule shows the resonance pattern, which is characteristic of the cyclopentadiene at 6.58 and 6.48 ppm, together with the resonances corresponding to the methyl and methylene protons in the organosilicon core. The integration of the resonances of the olefinic protons with the methyl resonance of the dendritic framework at −0.09 ppm confirms the complete peripheral tetrafunctionalization of the starting organosilicon dendrimer with cyclopentadiene.

The coordinating ability of these surface located cyclopentadienyl ligands was assessed via the reaction with octacarbonyldicobalt. Treatment of **15** with $Co_2(CO)_8$ in CH_2Cl_2 at reflux temperature in the presence of 1,3-cyclohexadiene and, afforded the desired brown-red derivative **16**. Support of the η^5-coordination of the cyclopentadienyl ligands to the $Co(CO)_2$ moieties included the characteristic transformations of the cyclopentadienyl ring resonances in the ^{1}H and ^{13}C NMR spectra as well as the presence of the two typical $\nu(CO)$ bands in the IR spectrum of **16**, which are in accord with the proposed structure. Thus in the ^{1}H NMR spectrum of **16**, the protons of the η^5-C_5H_4 ligands show the expected upfield shift (to 5.2 and 4.8 ppm) confirming the metal coordination to the four cyclopentadienyl ligands.

On the other hand, the reactivity of octacarbonyldicobalt toward Si–H functionalized organosilicon derivatives also offered a good chance to gain an easy synthetic access to a novel dendrimer **17**, which bears the organometallic moieties directly attached to the dendritic framework through silicon–metal s-bonds (Scheme 5). In this way, $Co_2(CO)_8$ reacts at room temperature in n-hexane with the Si–H functionalized dendrimer **G1H$_1$** in a process which involves hydrogen elimination and cleavage of the metal–metal bond in the starting carbonyl dimer to afford the tetrametallic dendritic macromolecule **17**. The IR spectrum shows the typical $\nu(CO)$ stretches at 2090, 2027 and 1992 cm^{-1} characteristic of the $Co-(CO)_4$ units and the total absence of the $\nu(Si-H)$ band at 2113 cm^{-1}, giving evidence of the complete silicon–cobalt attachment.

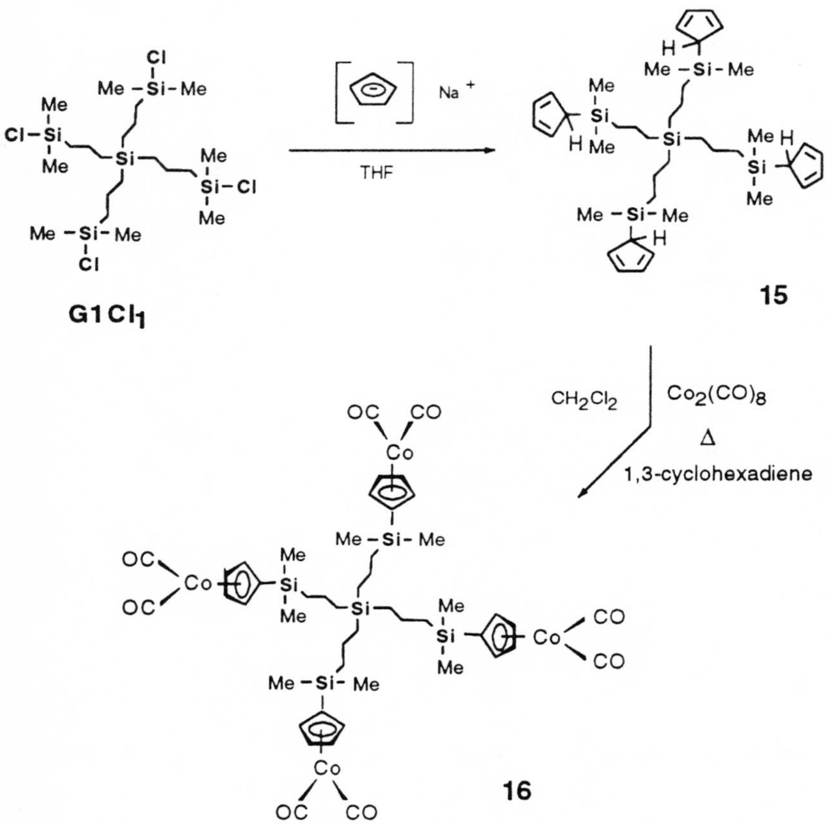

Scheme 4. Preparation of dendrimers with cyclopentadienyl groups bonded to the dendritic surface.

Likewise, an additional evaluation of the functionalization of the surface groups of the previously described SiCl-terminated dendrimer **G1Cl$_1$** was affected by its treatment with a THF solution of the carbonyl anion, Na$^+$[η^5-C$_5$H$_5$Fe(CO)$_2$]$^-$, affording the tetranuclear dendrimer **18** which contains at the periphery silicon–iron σ-bonds. The ^{29}Si NMR spectrum shows the two expected signals corresponding to the two different silicon atoms and interestingly, the external silicon directly bound to the iron atom appears shifted lowfield by about 40 ppm.

Work is in progress to advance all the above described syntheses to higher generations.

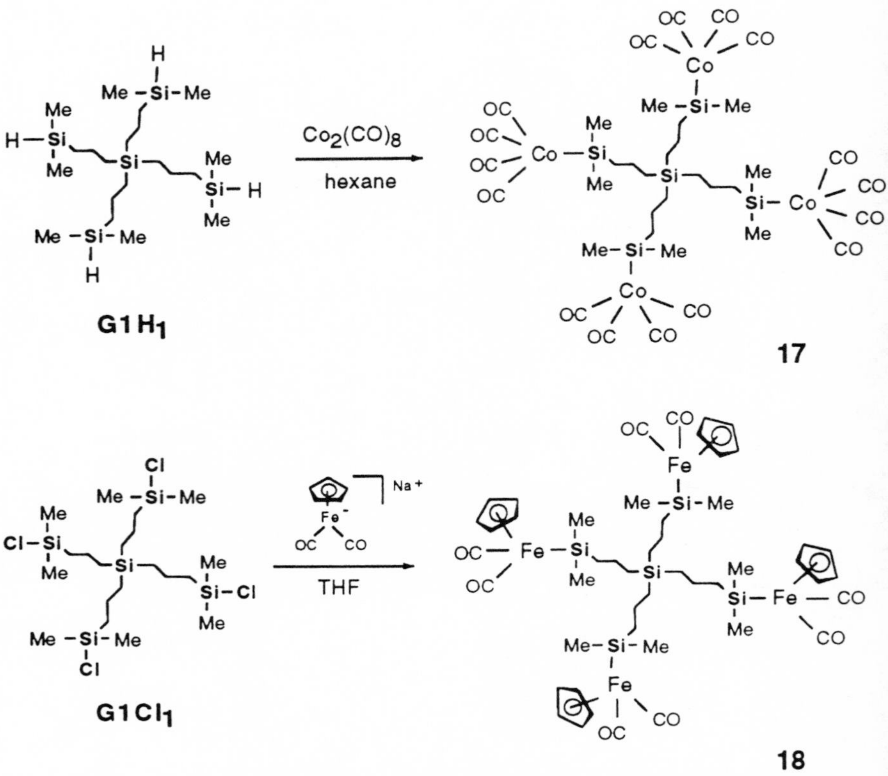

Scheme 5. Silane-based dendrimers containing at the periphery metal–silicon σ-bonds.

C. Organotransition Metal Silicon-Based Dendrimers Exhibiting Catalytic Activity

In 1994, van Koten and co-workers[42] reported remarkable silane-based dendrimers bearing catalytically active aryl–nickel sites precisely ordered onto the periphery of the dendritic framework. The synthesis of these interesting organometallic dendritic molecules involved the use of four-directional carbosilane core units and the diaminoaryl bromide moiety **20** as the precursor entity for the organometallic catalytic loci (see Scheme 6). For instance, reaction of the $SiMe_2Cl$-dodecafunctionalized silane **19** with the appropriate number of equivalents of **20** in CH_2Cl_2 in the presence of Et_3N, yielded the aryl bromide-substituted system **21**. This dendrimer was subsequently reacted with the zerovalent nickel derivative $Ni(PPh_3)_4$ in tetrahydrofuran at 60–70 °C in order to obtain the target oxidative addition product **22** possessing 12 discrete metal sites. In a similar way, a tetranu-

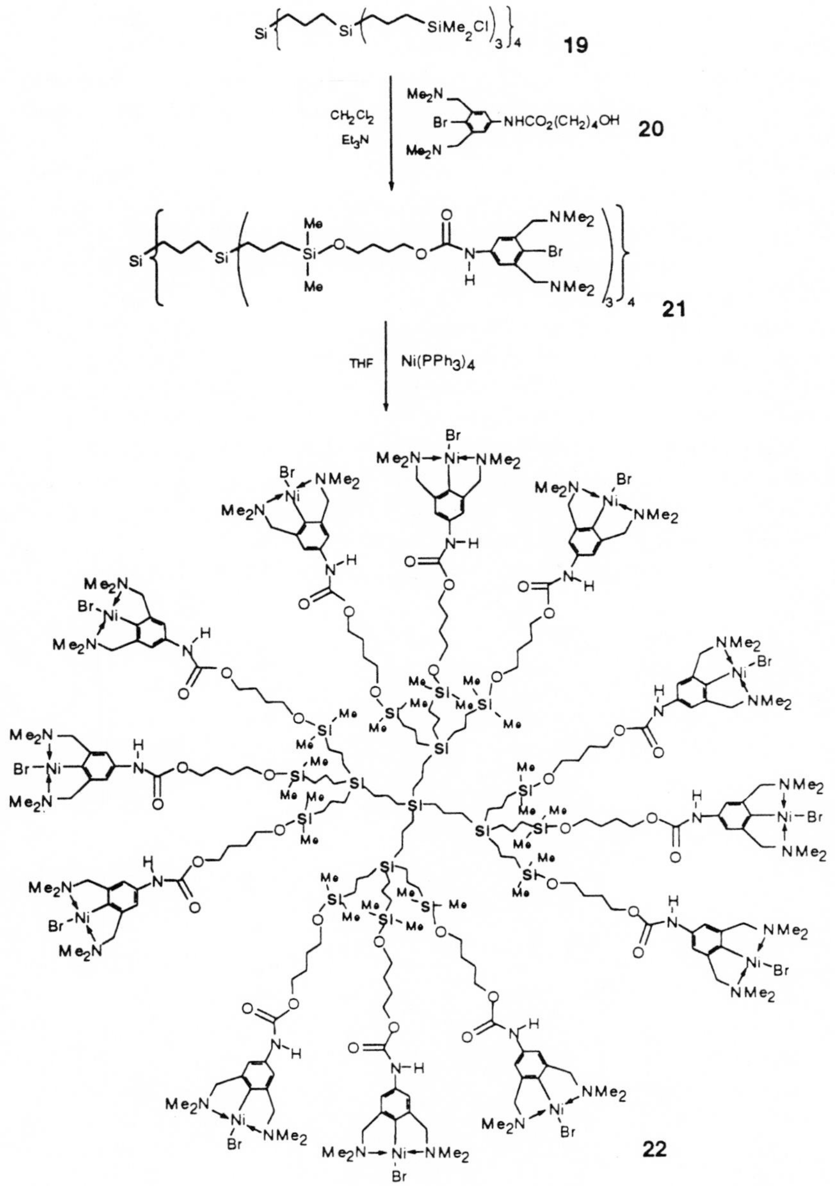

Scheme 6. van Koten's construction of a silane-based dendrimer functionalized at the periphery with arylnickel(II) catalytically active sites.

clear dendrimer was obtained starting from the corresponding SiMe$_2$Cl-tetrafunc-
tionalized core. Characterization of these nickel-containing dendrimers included
elemental analysis and ^{1}H and ^{13}C NMR spectroscopy.

Undoubtedly, the most notable feature of these new dendrimeric organometallic
molecules is their ability to act successfully as effective homogeneous catalysts for
the Kharasch addition reaction of polyhalogenoalkanes to olefinic C=C double
bonds. Indeed, they show catalytic activity and clean regiospecific formation of 1:1
addition products in a similar way to that observed in the mononuclear compounds.
Likewise, the nanoscopic size of these first examples of soluble dendritic catalysts
allows the separation of such macromolecules from the solution of the products by
ultrafiltration methods.

Organometallic dendrimers with catalytic activity represent a promising class of
materials for the future research in the field of catalysis. The combination of the
unique structural features of dendrimers with the rich catalytic chemistry, exhibited
by organometallic complexes, may allow the designed generation of novel efficient
catalysts with highly controlled architectures and precisely placed catalytic centers.
Likewise, as a result of the high degree of branching, large dendrimers adopt a
globular shape which assures the accessibility of all the catalytically active centers
to the reactive substrate. These new dendrimeric catalytic systems not only would
be expected to retain the advantages of homogeneous catalysts (high activity and

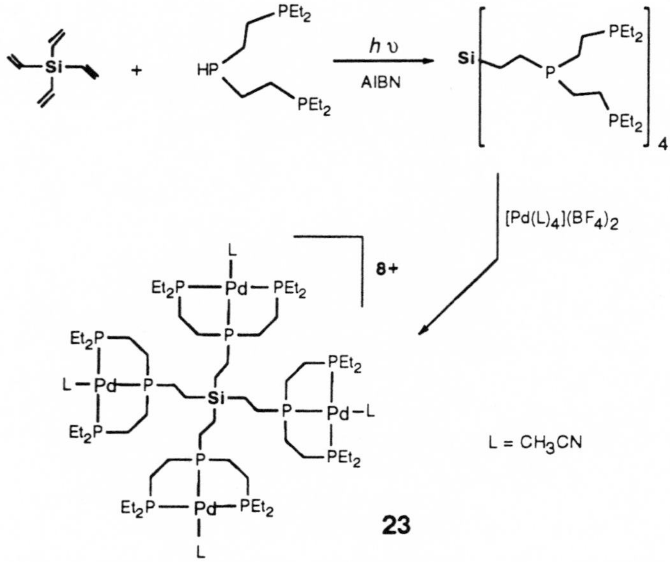

Scheme 7. DuBois's synthesis of a palladium containing organophosphine dendrimer
using a four-directional organosilicon core.

selectivity), but in addition, due to their nanoscopic size, would be easily separated from the resulting reaction–product mixtures.

Another interesting silane-based organophosphine dendrimer containing palladium which exhibits electrocatalytic activity was reported by DuBois et al.[43] The synthesis, illustrated in Scheme 7, was accomplished by free-radical addition of bis[(diethylphosphino)ethyl]phosphine to tetravinylsilane to yield a dendrimer possessing 12 phosphorus atoms characterized by ^{31}P NMR spectroscopy. Metallation of this dendrimer by reaction with $[Pd(CH_3CN)_4](BF_4)_2$ gives the dendritic macromolecule **23** which contains four well-separated palladium complexes bound to a central silicon atom. Electrochemical studies of **23**, in acidic dimethylformamide solution (0.05 M HBF_4) and in the presence of CO_2, showed an enhancement in the current of the reduction wave, indicating that this metallated dendrimer catalyzes the electrochemical reduction of CO_2 to CO. Interestingly, on the basis of the comparison of the catalytic activity for electrochemical reduction of CO_2 exhibited by **23** (in which the discrete palladium sites closely approximate typical homogeneous catalysts) with that observed for related palladium organophosphine dendrimers without organosilicon frameworks (in which no separation of the dendrimer into discrete palladium sites is inherent in their structures), the authors noted relationships between catalytic properties and dendrimer structure.

III. DENDRIMERS CONSTRUCTED FROM ORGANOMETALLIC ENTITIES

A wide variety of organic and inorganic dendrimers have been successfully prepared by using either *divergent*[2-5] or *convergent*[44,45] synthetic approaches, and most of them have been reviewed elsewhere in previous volumes of this series. In a similar way, a variety of organometallic dendrimers have been constructed by first preparing suitable organometallic moieties from which the buildup of the structure was accomplished by following either divergent or convergent growth synthetic strategies. For instance, a divergent approach to organometallic dendrimers involves growth from a polyfunctional organometallic initiator core where branching outward is accomplished via an increasing number of terminal branch transformations. Without a doubt, the main disadvantage of such divergent growth to organometallic dendrimers derives from the fact that a progressively large number of organometallic monomers have to react successfully with the reactive functional groups of the dendrimer surface. Therefore, as a result of the steric hindrance imposed by the steric-demanding organometallic units at higher generations it becomes extremely difficult to complete the reaction of all of the external functionalities of the dendrimer. For this reason, a critical aspect in the divergent strategy is the choice of a suitable organometallic initiator core in which an appropriate number of reactive sites are conveniently placed to assure that complete functionalization of the branches is accomplished easily and in relatively high yield. In the alternative convergent approach, the construction of organometallic dendrimers starts at what will ultimately become the outer surface of the dendrimer and progresses inward. Thus, this approach first requires the synthesis of conveniently functionalized, progressively larger organometallic *dendritic wedges* or *dendrons* in order to be subsequently attached around a polyfunctional central core, resulting in the final organometallic dendrimer. One attractive advantage of the convergent construction of dendrimers is that by coupling organometallic wedges of different nature to the same core molecule, *segment-block* or *layer-block organotransition heterometallic dendritic macromolecules* may be constructed.

A. Organometallic Entities as Initiator Cores: Divergent Approach

Astruc and co-workers were the first to report divergent approaches to the construction of organometallic molecular trees.[46–49] A representative example of their exciting synthetic strategy for the synthesis of multimetallic arborols is illustrated in Scheme 8.[47] The growing step is based on the fact that the π-complexation of aromatic compounds by electron-withdrawing cationic organo-transi-

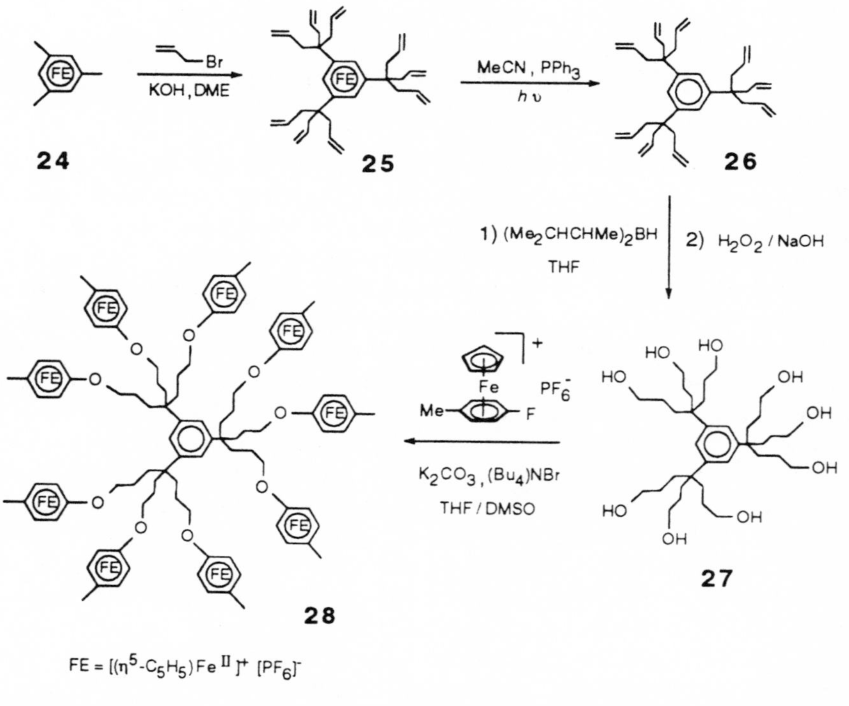

FE = [(η^5-C$_5$H$_5$)Fe^{II}]$^+$ [PF$_6$]$^-$

Scheme 8. Astruc's approach to the synthesis of a nonairon molecular tree using a three-directional organometallic core.

tion metal fragments enhances the acidity of methyl substituents in polymethyl hydrocarbon ligands, facilitating nucleophilic reactions. In this way, polyfunctionalization of methylated aromatics via their $(\eta^5\text{-}C_5H_5)M^+$ (M = Fe, Co) complexes has been successfully achieved.[46–49] In the synthesis shown in Scheme 8, the key starting organometallic entity employed as core molecule was the mesitylene derivative $[(\eta^6\text{-}C_6Me_3H_3)Fe(\eta^5\text{-}C_5H_5)]PF_6$ (**24**). Each methyl group in this monometallic initiator core was converted into a substituent with three allyl functionalized branches by one-pot nonaallylation of mesitylene with allyl bromide and KOH in dimethoxyethane. Clean demetallation of the resulting nonaolefin iron complex **25** by visible-light photolysis in acetonitrile in the presence of PPh_3 afforded the nonaallyl **26**, whose crystal structure was established by X-ray analysis. Regiospecific hydroboration with disiamylborane, followed by alkaline oxidation, afforded the nonaol **27**. Subsequently, the attachment of external organometallic units to this core molecule was accomplished by reaction of **27** with $[(\eta^6\text{-}p\text{-}MeC_6H_4F)Fe(\eta^5\text{-}C_5H_5)]PF_6$. The resulting nonairon molecular tree **28** bears external *p*-methyl substituents on the branches which potentially provide the possibility to iterate the starburst reaction at the level of a second generation. Examination of the electrochemical properties showed that the nonairon arborol **28** has a reversible reduction wave at -1.37 V vs. SCE in DMF/0.1 M tetrabutylammonium tetrafluoroborate corresponding to the Fe^{II} (18 e$^-$)/Fe^I (19 e$^-$) system and that 8 ± 1 electrons per molecule are transferred in the process. Therefore, the molecular tree **28** is a reservoir of nine electrons and 27 benzylic protons.

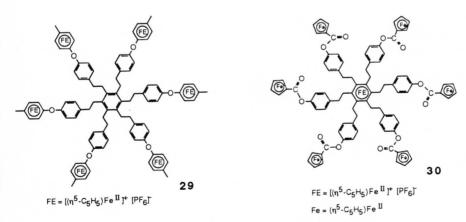

29

FE = $[(\eta^5\text{-}C_5H_5)Fe^{II}]^+$ $[PF_6]^-$

30

FE = $[(\eta^5\text{-}C_5H_5)Fe^{II}]^+$ $[PF_6]^-$

Fe = $(\eta^5\text{-}C_5H_5)Fe^{II}$

Figure 14. Hexaarm star-shaped iron sandwich complexes.

Likewise, also following this organoiron-mediated polysequential synthetic strategy, closely related star-shaped hexa- and heptanuclear iron sandwich molecules have been synthesized starting from a permethylated organometallic core molecule.[48,49] For instance, hexabenzylation of the initiator core $[(\eta^6\text{-}C_6Me_6)Fe(\eta^5\text{-}C_5H_5)]PF_6$, allowed the synthesis of hexaphenol species from which novel organometallic molecular trees such as **29** and **30**, with six equivalent redox centers attached at the end of the branches, were obtained (Figure 14).[48] All these organometallic molecular trees promise to be useful for multielectron redox catalysis.

On the other hand, Bunz and Enkelmann[50] used a different approach strategy for constructing polymetallic star-shaped tricarbonyl homo- and heterometallic complexes, such as **33** and **34** (Scheme 9). These were obtained by Pd-catalyzed coupling of the organometallic stannylacetylene and stannylbutadiyne derivatives **31** and **32**, respectively, to the key starting tetraiodo complex $(C_4I_4)Fe(CO)_3$. Interestingly, X-ray crystal structure analyses have been carried out for several of the synthesized multimetallic molecules.

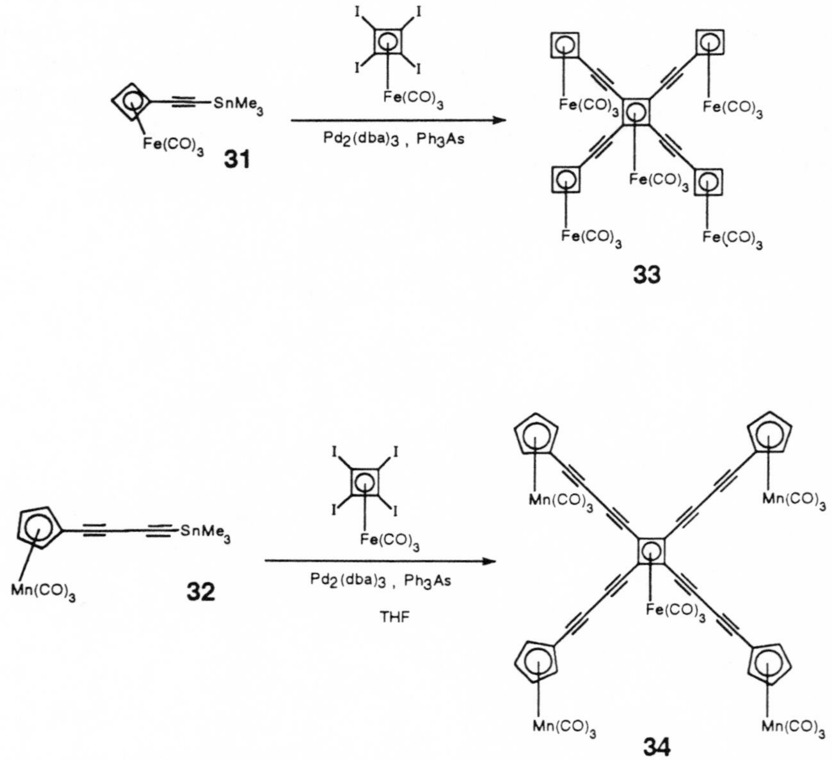

Scheme 9. Bunz's synthesis of star-shaped polymetallic tricarbonyl complexes.

B. Organometallic Entities as Dendrons: Convergent Approach

The first convergent approach to organometallic dendritic macromolecules was reported in 1993 by Liao and Moss,[51] resulting in dendrimers which contain at their peripheries 6 and 12 ruthenium–carbon *s*-bonds. In their strategy they have used the same convergent methodology developed by Fréchet and Hawker[44] for the synthesis of polyethereal dendrimers employing a three-directional trihydroxy core. The selected organometallic functionalized monomer was the bromoalkyl ruthenium derivative $(\eta^5\text{-}C_5H_5)Ru(CO)_2\{(CH_2)_3Br\}$ and the initial organic dendritic building block was 3,5-dihydroxybenzyl alcohol (see Scheme 10).

The construction of the organometallic dendron required two synthetic transformations. Selective alkylations of the phenolic hydroxyl groups in the presence of potassium carbonate and 18-crown-6 afforded the ether dimetallic derivative **35**. This first generation benzyl alcohol **35** was converted to the benzyl bromide **36** by

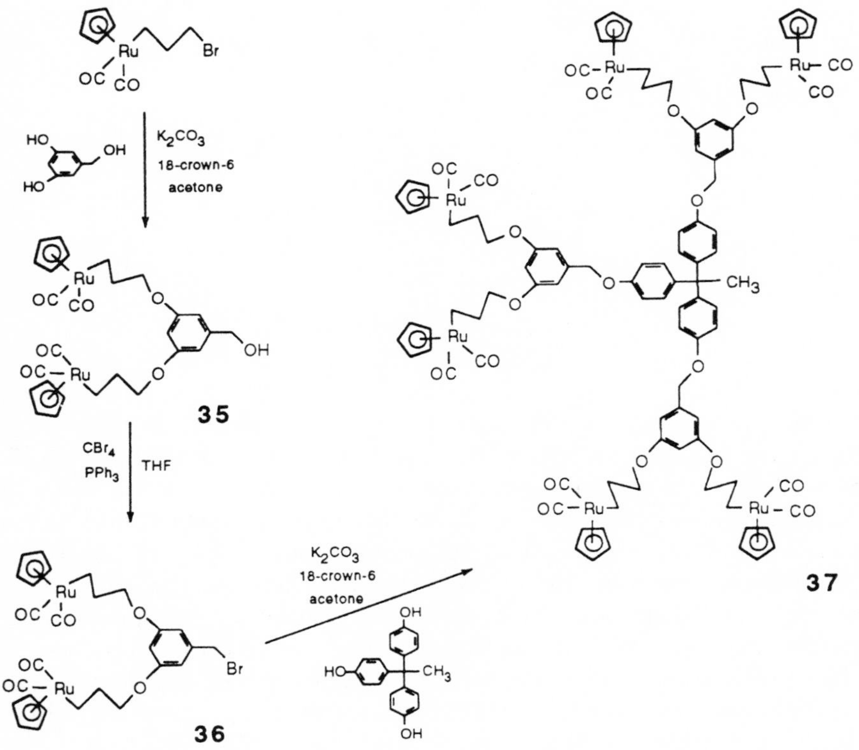

(continued)

Scheme 10. Moss's convergent construction of organometallic dendrimers containing ruthenium–carbon σ-bonds.

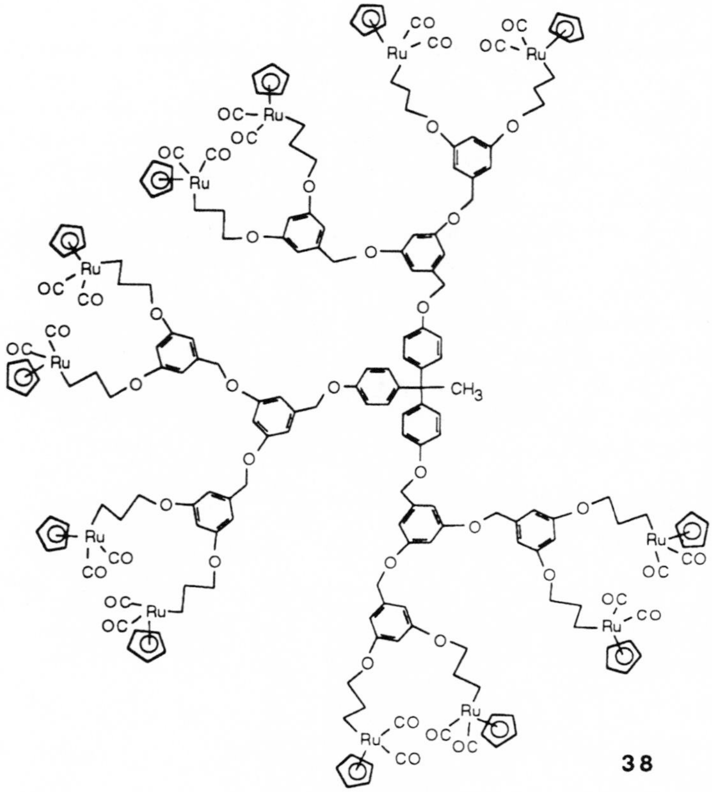

Scheme 10. (Continued)

treatment with an excess of PPh$_3$ and CBr$_4$. Dendritic wedge **36** was attached to a three-directional trisphenolic core affording the hexametallic dendritic system **37**. Similarly, by reacting a second-generation benzyl bromide dendritic fragment with the same trihydroxy core, dendrimer **38** with 12 dicarbonylcyclopentadienyl ruthenium moieties was prepared. Structural characterization included IR and ^{1}H and ^{13}C NMR spectroscopy as well as mass spectrometry. These were the first examples of organometallic dendrimers which contained organometallic moieties possessing metal–carbon s-bonds on the molecule surface. Very recently,[52] this stepwise convergent synthetic methodology has been extended to higher generations and, thus, a dendrimer containing 48 organoruthenium functional groups exclusively attached at the periphery of a poly(benzyl phenyl ether) dendritic structure has been prepared.

Achar and Puddephatt[53] reported a successful strategy for the synthesis of organoplatinum dendrimers containing up to 14 platinum atoms. These dendritic complexes were constructed convergently employing as alternating growing steps the two reactions shown in Scheme 11. First, oxidative addition of the C–Br bonds of the trifunctional 4,4′-bis(bromomethyl)-2,2′-bipyridine to the square-planar platinum(II) centers of [PtMe$_2$(bu$_2'$bpy)] (**39**) (bu$_2$bpy = 4,4′-di-*tert*-butyl-2,2′-bipyridine) gave the binuclear complex **40**, which contains two stable platinum(IV) centers and a free diimine functionality. Second, displacement of SMe$_2$ ligands from [Pt$_2$Me$_4$(m-SMe$_2$)$_2$] by the free bipyridine group of **40** gave the trinuclear

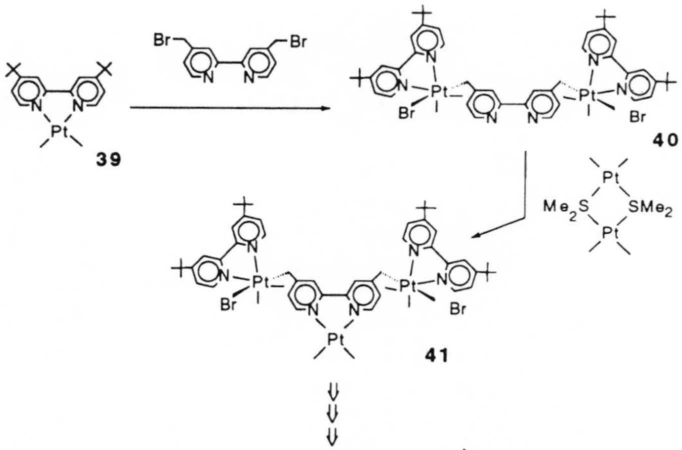

Scheme 11. Puddephatt's synthesis of organoplatinum dendrimers formed by oxidative addition.

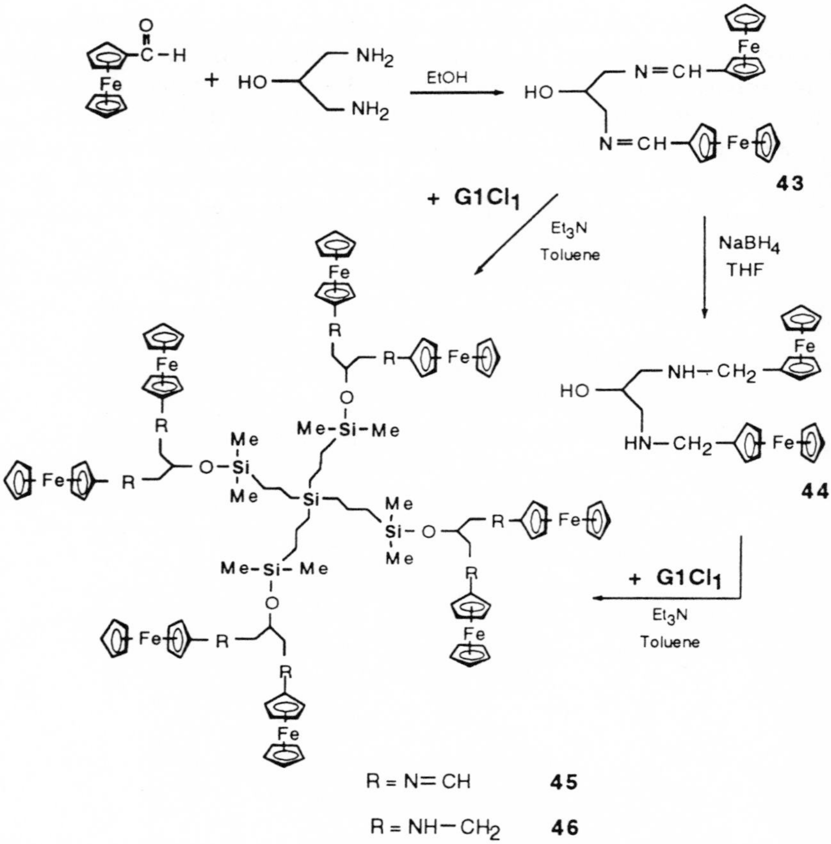

Scheme 12. Construction of silicon-based dendrimers using ferrocenyl-containing dendrons.

derivative **41**. After repetition of these alternating steps, the yellow dendrimer **42** was obtained in which steric hindrance at the reaction center precludes further growth to a dendrimer of higher nuclearity.

A modification of this method by using oxidative addition of 1,2,4,5-tetrakis(bromomethyl)benzene to a heptanuclear precursor with dimethylplatinum(II) centers allowed the synthesis of an organometallic dendrimer containing 28 platinum atoms.[54] In this dendritic organoplatinum complex, the metal centers are contained in every layer of the dendrimer except in the organic core. All the new organoplatinum dendrimers were characterized by UV-visible and ^{1}H NMR spectroscopies, gel permeation chromatography, and FAB mass spectrometry.

In our quest for new series of dendritic macromolecules possessing ferrocenyl moieties at predetermined sites into organosilicon frameworks, our group also convergently created a new family of organometallic silicon-based dendrimers, as shown in Scheme 12.[41] We selected 1,3-diaminopropan-2-ol as a two-directional building block for the construction of an organometallic dendron. Thus, the amine groups were condensed with ferrocenecarbaldehyde to afford cleanly, and in high yield, the ferrocenylimine dinuclear derivative **43** as an orange crystalline solid. This dendritic organometallic wedge was subsequently attached to the four-directional organosilicon core **G1Cl₁** giving the final octanuclear ferrocenyl-terminated dendrimer **45**. The IR spectrum of dendron **43**, as well as that of the octanuclear dendrimer **45**, exhibit the characteristic intense, sharp band at about 1640 cm^{-1} due to the $v(C{=}N)$ imine stretching peak. The ^1H and ^{13}C NMR spectra are also in agreement with the proposed structures.

On the other hand, the imino groups in the dinuclear **43** can be easily reduced using $LiAlH_4$ to give the diamino wedged dendron **44**, which was also subsequently reacted with **G1Cl₁** to afford the novel dendrimer **46** with eight independent ferrocenyl centers. IR spectra of both dendrimers are consistent with the reduction of the imino groups, showing the characteristic $v(NH)$ stretches in the region $3300–3100 \text{ cm}^{-1}$ and the total absence of the $v(CN)$ band. The most noteworthy features of these organometallic dendrimers are the presence of the NH functions, together with the particular cavities in the dendritic structures which promise to be of relevant importance in the electrochemical recognition of anionic guest species.

IV. FUTURE WORK AND PERSPECTIVES

Although the field of organometallic dendritic macromolecules is still in its infancy, it appears from the results described above that the research in the synthesis and characterization of such dendrimers within the last few years has provided a great deal of exciting and significant results.

Moreover, the opportunities for future progress in organometallic dendrimers are almost unlimited considering the diversity of synthetic routes, numerous variations in the topologies of the dendrimers, as well as the rich variety and the large number of organometallic entities and possible combinations of these which can be incorporated into dendritic structures. A challenging synthetic target is the construction of very large organometallic dendrimers. Indeed, work is currently in progress in order to extend some of the synthesis described above to higher generations. In this regard, the valuable information provided by molecular modeling studies is of crucial importance, not only to gain insight into the approximate dimensions, preferred conformations, and other structural features of these complex structures, but also to help in the design of new organized organometallic supermolecules.

Organometallic dendrimers have a key role to play in the future research in the field of catalysis. In fact, dendritic macromolecules of appropriate size, shape, density, and nuclearity, having organometallic entities precisely placed in accessible positions, could be designed in order to achieve specific interactions with reactive substrates for well-defined catalytic applications.

Preparation of photo-active and redox-active dendritic macromolecules, which undergo simultaneous multielectron transfer is also attractive. Considering the potential applications of these multimetallic dendrimers as electron transfer mediators in redox catalysis, photoinduced electron transfer, and molecular electronics, new interesting results can be awaited in the near future.

In addition, such redox-active organometallic dendrimers are interesting materials with which to modify electrode surfaces. Applications of these dendrimer modified electrodes in the fields of amperometric and potentiometric biosensors, molecular recognition, as well as in electrocatalysis and photoelectrochemistry, clearly represent interesting areas of future research.

Finally, the combination of dendrimers and organometallic entities as fundamental building blocks affords an opportunity to construct an infinite variety of organometallic starburst polymeric superstructures of nanoscopic, microscopic, and even macroscopic dimensions. These may represent a promising class of organometallic materials due to their specific properties, and potential applications as magnetic ceramic precursors, nonlinear optical materials, and liquid crystal devices in nanoscale technology.

In conclusion, considering all the attractive features of organometallic dendritic macromolecules, the continuous development of synthetic strategies for the construction of new families of such dendrimers, which will play a key role in the future research of supramolecular chemistry and material science, appears to be very promising.

ACKNOWLEDGMENTS

Our research on organometallic dendrimers is supported by the Dirección General de Investigación Científica y Técnica (DGICYT), Proyect 93/0287.

NOTE ADDED IN PROOF

The synthesis of carbosilane dendrimers with peripheral acetylenedicobalt hexacarbonyl substituents has been reported: Seyferth, D.; Kugita, T. *Organometallics* **1995**, *14*, 5362.

Organopalladium dendrimers prepared by self-assembly and controlled assembly methods have also been described: Huck, W. T. S.; van Veggel, F. C. J. M.; Kropman, B. L.; Blank, D. H. A.; Keim, E. G.; Smithers, M. M. A.; Reinhoudt, D. N. *J. Am. Chem. Soc.* **1995**, *177*, 8293, and Huck, W. T. S.; van Veggel, F. C. J. M.;

Reinhoudt, D. N. *Angew. Chem.* **1996**, *108*, 1304; *Angew. Chem., Int. Ed. Engl.* **1996**, *35*, 1213.

All the three families of silicon-based ferrocenyl dendrimers described in Section II.A of this chapter have been extended to the third generation containing 16 ferrocenyl moieties. In addition poly(propylene imine)-based organometallic dendrimers built up to the fifth generation containing 64 peripheral ferrocenyl moieties have been reported. Casado, C. M.; Cuadrado, I.; Morán, M.; Alonso, B.; Lobete, F.; Losada, J.; Barranco, M.; Moya, A.; García, B.; Ibisate, M. Presented at the 211th American Chemical Society National Meeting, New Orleans, LA, USA, March 1996, paper 178. Submitted for publication.

Convergent and divergent approaches to silicon-based organometallic dendrimers have been accomplished using key ferrocene derivatives such as $[(\eta^5-C_5H_4)Fe(\eta^5-C_5H_5)]_2SiMe(CH=CH_2)$ and $Fe[\eta^5-C_5H_4Si(Me)_2(CH=CH_2)]_2$ as dendron and initiator core, respectively. Cuadrado, I.; Casado, C. M.; Alonso, B.; Morán, M.; Losada, J.; Barranco, M.; García, B.; Ibisate, M. Presented at the XI International Symposium on Organosilicon Chemistry, Montpellier, France, September 1996, paper PB46.

REFERENCES AND NOTES

1. For recent reviews on dendritic macromolecules see: (a) *Advances in Dendritic Macromolecules*; Newkome, G. R., Ed.; JAI Press: Greenwich, CT, 1994, Vol. 1; 1995, Vol. 2; (b) Tomalia, D. A.; Durst, H. D. *Topics in Current Chemistry; Supramolecular Chemistry I-Directed Synthesis and Molecular Recognition*; Weber, E., Ed.; Springer-Verlag: Berlin, 1993, Vol. 165, p. 193; (c) Issberner, J.; Moors, R.; Vögtle, F. *Angew. Chem.* **1994**, *106*, 2507; *Angew. Chem. Int. Ed. Engl.* **1994**, *33*, 2413; (d) Fréchet, J. M. *Science* **1994**, *263*, 1710. (e) Tomalia, D. A. *Adv. Mater.* **1994**, *6*, 529; (f) Tomalia, D. A. *Aldrichimica Acta* **1993**, *26*, 91; (g) Newkome, G. R.; Moorefield, C. N.; Baker, G. R. *Aldrichimica Acta* **1992**, *25*, 31; (h) Mekelburger, H.-B.; Jaworek, W.; Vögtle, F. *Angew. Chem.* **1992**, *104*, 1609; *Angew. Chem. Int. Ed. Engl.* **1992**, *31*, 1571.

2. Buhleier, E.; Wehner, W.; Vögtle, F. *Synthesis* **1978**, 155.

3. Denkewalter, R. G.; Kolc, J. F.; Lukasavage, W. J. U.S. Patent 4 410 688, 1979.

4. Newkome, G. R.; Yao, Z.-Q.; Baker, G. R.; Gupta, V. K. *J. Org. Chem.* **1985**, *50*, 2003.

5. Tomalia, D. A.; Baker, H.; Dewald, J.; Hall, M.; Kallos, G.; Martin, S.; Roeck, J.; Ryder, J.; Smith, P. *Polym. J.* **1985**, *17*, 117.

6. See for example: (a) Miller, L. L.; Hashimoto, T.; Tabakovic, I.; Swanson, D. R.; Tomalia, D. A. *Chem. Mater.* **1995**, *7*, 9; (b) Sournies, F.; Crasnier, F.; Graffeuil, M.; Faucher, J.-P.; Lahana, R.; Labarre, M.-C.; Labarre, J.-F. *Angew. Chem.* **1995**, *107*, 610; *Angew. Chem. Int. Ed. Engl.* **1995**, *34*, 578; (c) Galliot, C.; Prévoté, D.; Caminade, A.-M.; Majoral, J.-P. *J. Am. Chem. Soc.* **1995**, *117*, 5470; (d) Bryce, M. R.; Devonport, W.; Moore, A. J. *Angew. Chem.* **1994**, *106*, 1862; *Angew. Chem. Int. Ed. Engl.* **1994**, *33*, 1716; (e) Launay, N.; Caminade, A.-M.; Lahana, R.; Majoral, J.-P. *Angew. Chem.* **1994**, *106*, 1682; *Angew. Chem. Int. Ed. Engl.* **1994**, *33*, 1598; (f) Ottaviani, M. F.; Bossmann, S.; Turro, N. J.; Tomalia, D. A. *J. Am. Chem. Soc.* **1994**, *116*, 661.

7. Moors, R.; Vögtle, F. *Chem. Ber.* **1993**, *126*, 2133.

8. (a) Serroni, S.; Denti, G.; Campagna, S.; Juris, A.; Ciano, M.; Balzani, V. *Angew. Chem.* **1992**, *104*, 1540; *Angew. Chem., Int. Ed. Engl.* **1992**, *31*, 1493; (b) Denti, G.; Campagna, S.; Serroni, S.; Ciano, M.; Balzani, V. *J. Am. Chem. Soc.* **1992**, *114*, 2944; (c) See also Chapter 2 in this book.

9. Jin, R. H.; Aida, T.; Inoue, S. *J. Chem. Soc., Chem. Commun.* **1993**, 1260.

10. Dandliker, P. J.; Diederich, F.; Gross, M.; Knobler, C. B.; Louati, A.; Sanford, E. M. *Angew. Chem.* **1994**, *106*, 1821; *Angew. Chem., Int. Ed. Engl.* **1994**, *33*, 1739.

11. Newkome, G. R.; Moorefield, C. N. *Polym. Prepr., Am. Chem. Soc. Div. Polym. Chem.* **1993**, *34*, 75.

12. Newkome, G. R.; Cardullo, F.; Constable, E. C.; Moorefield, C. N.; Thompson, A. M. W. C. *J. Chem. Soc., Chem. Commun.* **1993**, 925.

13. For a general review of organometallic polymers see: (a) *Inorganic and Organometallic Polymers II. Advanced Materials and Intermediates*; Wisian-Neilson, P.; Allcock, H. R.; Wynne, K. J., Eds.; ACS Symposium Series 572; American Chemical Society: Washington, DC, 1994; (b) *Inorganic and Metal-Containing Polymeric Materials*; Sheats, J. E.; Carraher, C. E., Jr.; Pittman, C. U., Jr.; Zeldin, M., Currell, B., Eds.; Plenum Press: New York, 1990; (c) *Inorganic Polymers*; Mark, J. E.; Allcock, H. R.; West, R., Eds.; Prentice-Hall: Englewood Cliffs, NJ, 1992; (d) *Inorganic and Organometallic Polymers*; Zeldin, M.; Wynne, K. J.; Allcock, H. R., Eds.; ACS Symposium Series 360; American Chemical Society: Washington, DC, 1988. (e) *Metal-Containing Polymeric Systems*; Sheats, J. E.; Carraher, C. E., Jr.; Pittman, C. U., Jr., Eds.; Plenum Press: New York, 1985.

14. (a) Allcock, H. R. *Adv. Mater.* **1994**, *6*, 106; (b) Peuckert, M.; Vaahs, T.; Brück, M. *Adv. Mater.* **1990**, *2*, 398; (c) Sergeev, V. A.; Vdovina, L. I. *Organomet. Chem. USSR* **1989**, *2*, 77.

15. Casado, C. M.; Cuadrado, I; Morán, M.; Alonso, B.; Lobete, P.; Losada, J. *Organometallics* **1995**, *14*, 2618.

16. Casado, C. M.; Morán, M.; Losada J.; Cuadrado, I. *Inorg. Chem.* **1995**, *34*, 1668.

17. Morán, M.; Casado, C. M.; Cuadrado, I.; Losada, J. *Organometallics* **1993**, *12*, 4237.

18. Recent reviews of the main areas of research on ferrocenes include the following: (a) *Ferrocenes*; Togni, A.; Hayashi, T., Eds.; VCH: Weinheim, 1995. See also (b) Foucher, D. A.; Ziembinski, R.; Rulkens, R.; Nelson, J.; Manners, I., Chapter 33 in Ref. 13 (a); (c) Wright, M. E.; Cochran, B. B.; Toplikar, E. G.; Lackritz, H. S.; Kerney, J. T., Chapter 34 in Ref. 13 (a); (d) Manners, I. *Adv. Organomet. Chem.* **1995**, *37*, 131.

19. For recent works on ferrocene polymers used for electrode surface modification, see: (a) Nguyen, M. T.; Diaz, A. F.; Dement'ev, V. V.; Pannell, K. H. *Chem. Mater.* **1994**, *6*, 952; (b) Inagaki, T.; Lee, H. S.; Skotheim, T. A.; Okamoto, Y. *J. Chem. Soc., Chem. Commun.* **1989**, 1181; (c) Albagly, D.; Bazan, G.; Wrighton, M. S.; Schrock, R. R., *J. Am. Chem. Soc.* **1992**, *114*, 4150; (d) Crumbliss, A. L.; Cooke, D.; Castillo, J.; Wisian-Neilson, P. *Inorg. Chem.* **1993**, *32*, 6088; (e) Saraceno, R. A.; Riding, G. H.; Allcock, H. R.; Ewing, A. G. *J. Am. Chem. Soc.* **1988**, *110*, 980; **1988**, *110*, 7254; (f) See also Refs. 15–17.

20. See for example (a) Wang, C.-L.; Mulchandani, A. *Anal. Chem.* **1995**, *67*, 1109; (b) Hale, P. D.; Lee, H. S.; Okamoto, Y. *Anal. Lett.* **1993**, *26*, 1; (c) Hale, P. D.; Lan, H. L.; Boguslavsky, L. I.; Karan, H. I.; Okamoto, Y.; Skotheim, T. A. *Anal. Chim. Acta* **1991**, *251*, 121; (d) Hale, P. D.; Inagaki, T.; Karan, H. I.; Okamoto, Y.; Skotheim, T. A. *J. Am. Chem. Soc.* **1989**, *111*, 3482.

21. Kittlesen, G. P.; White, H. S.; Wrighton, M. S. *J. Am. Chem. Soc.* **1985**, *107*, 7373.

22. Wright, M. E.; Cochran, B. B.; Toplikar, E. G.; Lackritz, H. S.; Kerney, J. T., in Ref. 13 (a), Chapter 34, and references therein.

23. Alonso, B.; Cuadrado, I.; Morán, M.; Losada, J. *J. Chem. Soc., Chem. Commun.* **1994**, 2575.

24. van der Made, A. W.; van Leeuwen, P. W. N. M. *J. Chem. Soc., Chem. Commun.* **1992**, 1400.

25. Zhou, L.-L.; Roovers, J. *Macromolecules* **1993**, *26*, 963.

26. Seyferth, D.; Son, D. Y.; Rheingold, A. L.; Ostrander, R. L. *Organometallics* **1994**, *13*, 2682.

27. Gonsalves, K. E.; Lenz, R. W.; Rausch, M. D. *Appl. Organomet. Chem.* **1987**, *1*, 81; and references therein.

28. Alonso, B; Morán, M; Casado, C. M.; Lobete, P.; Losada, J.; Cuadrado, I. *Chem. Mater.* **1995**, *7*, 1440.

29. (a) Sohn, Y. S.; Hendrickson, D. N.; Gray, H. B. *J. Am. Chem. Soc.* **1971**, *93*, 3603; (b) Nguyen, T.; Diaz, A. F.; Dement'ev, V. V.; Pannell, K. H. *Chem. Mater.* **1993**, *5*, 1389; (c) Pannell, K. H.;

Sharma, H. K. *J. Organomet. Chem.* **1993**, 450, 193; (d) Duggan, D. M.; Hendrickson, D. N. *Inorg. Chem.* **1975**, *14*, 955; (e) Cowan, D. O.; Candela, G. A.; Kaufman, F. *J. Am. Chem. Soc.* **1971**, *93:16*, 3889.

30. Geiger, W. E. *J. Organomet. Chem. Libr.* **1990**, *22*, 142.

31. Bard, A. J.; Faulkner, L. R. *Electrochemical Methods*; Wiley: New York, 1980.

32. Alonso, B.; Cuadrado, I.; Casado, C. M.; Morán, M.; Losada, J., manuscript in preparation.

33. (a) Abruña, H. D. In *Electroresponsive Molecular and Polymeric Systems*; Skotheim, T. A., Ed.; Dekker: New York, 1988, Vol 1. p. 97; (b) Murray, R. W. In *Molecular Design of Electrode Surfaces*; Murray, R. W., Ed.; Techniques of Chemistry XXII; Wiley: New York, 1992, p. 1.

34. See for example: (a) Beer, P. D. *Adv. Inorg. Chem.* **1992**, *39*, 79; (b) Beer, P. D.; Chen, Z.; Goulden, A. J.; Graydon, A.; Stokes, S. E.; Wear, T. *J. Chem. Soc., Chem. Commun.* **1993**, 1834; (c) Dietrich, B. *Pure Appl. Chem.* **1993**, *65*, 1457; (d) Verboom, W.; Rudkevich, D. M.; Reinhoudt, D. N. *Pure Appl. Chem.* **1994**, *66*, 679; (e) Beer, P. D.; Stokes, S. E. *Polyhedron* **1995**, *14*, 873; (f) Beer, P. D.; Drew, M. G. B.; Graydon, A. R.; Smith, D. K.; Stokes, S. E. *J. Chem. Soc., Dalton Trans.* **1995**, 403.

35. See for example: (a) Kalcher, K.; Kauffmann, J. M.; Wang, J.; Svancara, I.; Vytras, K.; Neuhold, C.; Yang, Z. *Electroanalysis* **1995**, *7*, 5; (b) Gilmartin, M.; Hart, J. P. *Analyst* **1995**, *120*, 1029; (c) Bartlett, P. N.; Cooper, J. M. *J. Electroanal. Chem.*, **1993**, *362*, 1.

36. Losada, J.; Cuadrado, I.; Morán, M.; Casado, C. M.; Alonso, B.; Barranco, M. *Anal. Chim. Acta*, in press.

37. Morán, M.; Cuadrado, I.; Pascual, C.; Casado, C. M.; Losada, J. *Organometallics* **1992**, *11*, 1210.

38. Davis, R.; Kane-Maguire, L. A. P. *Comprehensive Organometallic Chemistry*; Wilkinson, G.; Stone, F. G. A.; Abel, E. W., Eds.; Pergamon: Oxford, England, 1982, Vol. 3, p. 1001; and references therein.

39. Lobete, F.; Cuadrado, I.; Casado, C. M.; Alonso, B.; Morán, M.; Losada, J. *J. Organomet. Chem.* **1996**, *509*, 109.

40. (a) Janiak, C.; Schumann, H. *Adv. Organomet. Chem.* **1991**, *33*, 291; (b) Okuda, J. *Top. Curr. Chem.* **1992**, *160*, 97; (c) Jutzi, P. *Chem. Rev.* **1986**, *86*, 983.

41. Cuadrado, I.; Morán, M.; Moya, A.; Casado, C. M.; Barranco, M.; Alonso, B. *Inorg. Chim. Acta*, in press.

42. Knapen, J. W. J.; van der Made, A. W.; de Wilde, J. C.; van Leeuwen, P. W. N. M.; Wijkens, P.; Grove, D. M.; van Koten, G. *Nature* **1994**, *372*, 659.

43. Miedaner, A.; Curtis, C. J.; Barkley, R. M.; DuBois, D. L. *Inorg. Chem.* **1994**, *33*, 5482.

44. Hawker, C. J.; Frechet, J. M. J. *J. Chem. Soc., Chem. Commun.* **1990**, 1010.

45. Miller, T. M.; Neenan, T. X. *Chem. Mater.* **1990**, *2*, 346.

46. (a) Moulines, F.; Gloaguen, B.; Astruc, D. *Angew. Chem.* **1992**, *104*, 452; *Angew. Chem. Int. Ed. Engl.* **1992**, *31*, 458; (b) Cloutet, E.; Fillaut, J.-L.; Gnanou, Y.; Astruc, D. *J. Chem. Soc., Chem. Commun.* **1994**, 2433; (c) Astruc, D. *Top. Curr. Chem.* **1991**, *160*, 47.

47. Moulines, F.; Djakovitch, L.; Boese, R.; Gloaguen, B.; Thiel, W.; Fillaut, J.-L.; Delville, M.-H.; Astruc, D. *Angew. Chem.* **1993**, *105*, 1132; *Angew. Chem. Int. Ed. Engl.* **1993**, *32*, 1075.

48. Fillaut, J.-L.; Linares, J.; Astruc, D. *Angew. Chem.* **1994**, *106*, 2540; *Angew. Chem. Int. Ed. Engl.* **1994**, *33*, 2460.

49. Fillaut, J.-L.; Astruc, D. *J. Chem. Soc., Chem. Commun.* **1993**, 1320.

50. Bunz, U. H. F; Enkelmann, V. *Organometallics* **1994**, *13*, 3823.

51. Liao, Y.-H.; Moss, J. R. *J. Chem. Soc., Chem. Commun.* **1993**, 1774.

52. Liao, Y.-H.; Moss, J. R. *Organometallics* **1995**, *14*, 2130.

53. Achar, S.; Puddephatt, R. J. *Angew. Chem.* **1994**, *106*, 895; *Angew. Chem. Int. Ed. Engl.* **1994**, *33*, 847.

54. (a) Achar, S.; Puddephatt, R. J. *J. Chem. Soc., Chem. Commun.* **1994**, 1895; (b) Achar, S.; Vittal, J. J.; Puddephatt, R. J. *Organometallics* **1996**, *15*, 43.

INDEX

Advances in Dendritic Macromolecules

Edited by **George R. Newkome,** *Department of Chemistry, University of South Florida*

Volume 1, 1994, 198 pp. $109.50
ISBN 1-55938-696-7

Volume 2, 1995, 204 pp. $109.50
ISBN 1-55938-939-7

JAI PRESS INC.
55 Old Post Road No. 2 - P.O. Box 1678
Greenwich, Connecticut 06836-1678
Tel: (203) 661- 7602   Fax: (203) 661-0792

Advances in
Supramolecular Chemistry

Edited by **George W. Gokel**,
Department of Chemistry, University of Miami

**J A I
P R E S S**

Advances in Theoretically Interesting Molecules

Edited by **Randolph P. Thummel,** *Department of Chemistry, University of Houston*

Volume 1, 1989, 467 pp. $109.50
ISBN 0-89232-869-X

CONTENTS: Introduction to the Series: An Editor's Foreword, *Albert Padwa.* Preface, *Randolph P. Thummel.* Isobenzofurans, *Bruce Rickborn.* Dihydropyrenes: Bridged [14] Annulenes Par Excellence. A Comparison with Other Bridged Annulenes, *Richard H. Mitchell.* [l.m.n.] Hericenes and Related Exocyclic Polyenes, *Pierre Vogel.* The Chemistry of Pentacyclo $[5.4.0.0^{2,6}.0^{3,10}.0^{5,9}]$ Undecane (PCUD) and Related Systems, *Alan P. Marchand.* Cyclic Cumulenes, *Richard P. Johnson.* Author Index. Subject Index.

Volume 2, 1992, 223 pp. $109.50
ISBN 0-89232-953-X

CONTENTS: List of Contributors. Introduction to the Series: An Editor's Forward, *Albert Padwa.* Preface, *Randolph P. Thummel.* Cyclooctatetraenes: Conformational and π-Electronic Dynamics Within Polyolefinic [8] Annulene Frameworks, *Leo A. Paquette.* A Compilation and Analysis of Structural Data of Distorted Bridgehead Olefins and Amides, *Timothy G. Lease and Kenneth J. Shea.* Nonplanarity and Aromaticity in Polycyclic Benzenoid Hydrocarbons, *William C. Herndon and Paul C. Nowak.* The Dewar Furan Story, *Ronald N. Warrener.* Author Index. Subject Index.

Volume 3, 1995, 316 pp. $109.50
ISBN 1-55938-698-3

CONTENTS: Preface, *Randolph P. Thummel.* Polynuclear Aromatic Hydrocarbons with Curved Surfaces: Hydrocarbons Possessing Carbon Frameworks Related to Buckminsterfullerene, *Peter W. Rabideau and Andrzej Sygula.* Chemistry of Cycloproparenes, *Paul Muller.* A Tale of Three Cities: Planar Dehydro [8] Annulenes and Their Reverberations, *Henry N.C. Wong.* Infrared Spectroscopy of Highly Reactive Organic Species: The Identification of Unstable Molecules and Reactive Intermediates Using AB Initio Calculated Infrared Spectra, *B. Andes Hess, Jr. and Lidia Smentek-Mielczarek.* The Mills-Nixon Effect?, *Natia L. Frank and Jay S. Siegel.* Radical Cations of Cyclopropane Systems-Conjugation and Homoconjugation with Alkene Functions, *Heinz D. Roth.* Subject Index. Author Index.

J A I   P R E S S